Astronomers' Universe

For other titles published in this series, go to
www.springer.com/series/6960

David A. J. Seargent

Weird Astronomy

Tales of Unusual, Bizarre, and Other
Hard to Explain Observations

Springer

David A. J. Seargent
2261 The Entrance
Australia

520
5439w

ISBN 978-1-4419-6423-6   e-ISBN 978-1-4419-6424-3
DOI 10.1007/978-1-4419-6424-3
Springer New York Dordrecht Heidelberg London

Library of Congress Control Number: 2010935188

Printed on acid-free paper

Springer is part of Springer Science+Business Media (www.springer.com)

For Meg

Preface

Astronomy has a recorded history longer than any other physical science. True, its juvenile years were spent as a co-joined twin of its more dubious sibling, astrology, and its musings were often not what we in this supposedly more enlightened epoch would term "scientific." But we must always bear in mind that, for the ancients, astrological speculations were regarded with the same solemnity that we now reserve for the most profound of cosmological investigations.

Needless to say, such an ancient and venerable science has accrued around itself a rather mixed collection of interesting, sometimes puzzling, at times amusing, and on occasion downright bizarre accounts and anecdotes, ranging from interesting little tidbits of human interest to genuinely puzzling and anomalous observations. Lying beyond the mainstream of their topic, astronomical textbooks seldom mention these, but it is precisely such anecdotes and stories that make astronomy a living and human endeavor, as well as giving it a lighter side.

This "lighter side" – this fringe of anecdotes, oddities, factual trivia, and titillating tales – is what this book is all about. The title "Weird Astronomy" implies a wide range beyond the more or less staid mainstream of the topic. It is not necessarily "bad" astronomy. True, some examples of truly bad astronomy have been included here. But much "weird" astronomy is still "good" astronomy. It just does not fit well with what is official. A bright object near the Sun, a fast-moving blob of nebulosity in the night sky, a meteor that makes a "swishing" sound as it crosses the sky, lightning seen on the Moon or changing patterns on Mars, cell-like structures found in meteorites, or apparent signs of life on the Martian surface all make grist for the mill of this book.

Then there are the tales of difficult observations, naked-eye sightings of Jupiter's brighter moons, a quasar spied through a small backyard telescope, even an observation of sunspots made by peering up through the eye of a tropical storm. Such tales and much more await you in this book.

But we are not just collecting anecdotes here. There are activities for you, the reader, as well. Of course, we can't replicate lightning-like events on the Moon, nor is it likely that you will encounter repeat performances of most of the transitory events mentioned in these pages. And, of course, nobody observes the planet Vulcan these days!

Still, the way is open for you, the reader, to join in the fun and maybe even make some real contributions to science while so doing. Several "projects" have been included in these chapters which, it is hoped, will let you participate in at least some of the things about which you are reading. These projects are of varying degrees of complexity and will require different levels of experience. Those that require more experience or observing skills are marked with an asterisk (*). Some of these, if seriously and carefully pursued, may yield scientifically valuable data and, for these, basic directions are given for reporting your results to the interested parties.

In short then, if you are only interested in reading about the latest advances in planetary science, the most recent cosmological controversies, or the up-to-date count of extra-solar planets, you might want to stop right now.

But if the odd, the interesting, the peculiar, and – yes – the slightly weird attract you, and if you experience the thrill of a little practical observing as well ... read right on!

About the Author

David Seargent has a doctorate degree in philosophy from the University of Newcastle in Australia, is a former part-time teacher there, an amateur astronomer, and now a full-time writer. He has written several books and papers on comets, including one for Springer called Greatest Comets in History (2009). He was formerly a contributing editor on comets to *Sky and Space* magazine and is currently the author of a regular comets' column for *Australian Sky & Telescope*. Dr. Seargent formed the University Astronomical Society when a student at the University of Newcastle and later had the distinction of discovering a new comet, named for him, C/1978 T1 (Seargent)

Acknowledgements

This book is loosely based on a series of regular articles published over a number of years in the magazine *Southern Astronomy* (later *Sky & Space*) and would never have come to fruition without the early support of the publisher and editor of this journal. So a belated "thank you" must go to all involved, especially to editor Jon Nally for his encouragement and assistance way back then.

As the book began taking shape from these scattered seeds, many people contributed in one way or another to its evolution. I would especially like to thank Gordon Garradd of Siding Spring Observatory, for permission to use his comet and Venus transit images, Paul Camilleri, G. Sostero, V. Gonano, and E. Guido for their images of near-Earth asteroids, Dr. Peter Jenniskens and colleagues Mauwia Shaddad and Mohamed Elhassan Abdelatif Mahir for permission to reproduce their image of the spectacular dust trail of the Almahata Sitta Meteorite, Martin Elsasser for permission to reproduce his images of very (very!) thin lunar crescents, Emil Neata for permission to reproduce his finder chart of quasar 3C 273, and Michael Mullenniex of Malin Space Science Systems, Inc., for granting permission to reproduce images. I would also like to thank Drs. William Hartmann and Mark Garlic for allowing me to include samples of their fine space art within the pages of this book.

Many thanks also to Mr. John Watson and Ms. Maury Solomon of Springer Publishing for their advice and encouragement during the preparation of this book.

And of course, thank you dear reader for purchasing this book. May you be amply rewarded in your journey through the byways of our subject!

David A. J. Seargent
The Entrance
November 2009

Contents

1. Our Weird Moon

Without doubt, the Sun and Moon are the most recognized objects in the sky. The Moon is also our nearest celestial neighbor, excepting the occasional small asteroid that briefly skims past our world at disturbingly close range.

Our natural satellite is also the object that most nascent astronomers aim their first telescopes toward.

How many active amateur astronomers remember their first telescopic view of a celestial object? For how many was this object the Moon?

For most, we would venture to guess. Many of these amateurs, and many professionals as well, continued to make the Moon their chief astronomical interest.

The Moon also remains the only cosmic body besides our home planet that humankind has, albeit briefly, visited in person. Without it, we may never have ventured into space at all. (Actually, as we will discuss later, we may not have had that opportunity if the Moon did not exist. There are sound reasons to think that our very existence depends upon the presence of a large Moon!)

Surely, after years of such scrutiny and familiarity, the Moon holds no unaccountable weirdness!

Well, actually, that is not quite right …

Once Upon a Canterbury Evening

The date is June 18 in the year AD 1178. You are one of a small group of monks from Canterbury in England, quietly enjoying the balmy summer evening. In the western sky hangs a crescent Moon, bright and splendid in the fading twilight. The scene is one of peace and serenity.

D.A.J. Seargent, *Weird Astronomy*, Astronomers' Universe,
DOI 10.1007/978-1-4419-6424-3_1, © Springer Science+Business Media, LLC 2011

Then it happens!

From midway between the horns of the crescent Moon, an eruption of fire bursts forth like a flaming torch. What appear to be flames of fire, hot coals, and sparks spew outward from the Moon into the surrounding sky. The very Moon itself seems to writhe in pain, throbbing and twisting like a wounded serpent before once again regaining its composure. This phenomenon is repeated at least a dozen times in quick succession; the flame randomly taking a variety of twisting shapes before regaining the appearance of a torch. After the event, the Moon's appearance darkens, as if shrouded by a sort of cloud or mist.

Such was the event recorded by Gervase of Canterbury and his fellow monks on that long ago summer's evening. What other thoughts and reactions crossed their minds was not said, but we could imagine a certain terror gripping them as they watched this strange transformation of our familiar Moon.

But what did they actually see? What could possibly explain such an extreme sight?

A possible clue emerged with early mapping of the far side of the Moon (the side forever not visible from Earth) by unmanned Soviet probes during the opening years of the space race. Images sent back from these craft revealed a hitherto unknown and obviously very fresh crater located just beyond the boundary between the Earth-facing and "hidden" hemispheres of the Moon. This crater (subsequently named Giordano Bruno in honor of the controversial philosopher) is situated such that when the Moon is a crescent shortly after new, it is (from Earth's perspective) located roughly midway between the crescent's horns, but just over the lunar rim. In other words, it seems to be very close to the site of the Canterbury monks' "flaming torch." The first person to draw attention to this was astronomer Jack Hartung, who in 1976 published his thoughts on the matter in the journal *Meteoritics*.

Moreover, sensitive seismic equipment left on the Moon by *Apollo* astronauts revealed a curious fact about our satellite's recent seismic history. It was quickly discovered that the Moon reacts to impacts in a very different way than Earth does. By deliberately crashing a disused lunar module onto the Moon's surface, the *Apollo* scientific team found that an impact sets the Moon seismically "ringing" and that this effect only slowly damps down over time. This was further confirmed as the seismographs continued to pick up meteorite impacts on the lunar surface.

Curiously though, there seemed to be a gentle background ringing, over and above that of the frequent meteorite hits. This was thought to be the fading embers of a violent disturbance at some time in the more distant past, presumably caused by the impact of a really big meteorite or small asteroid. Moreover, further sensitive monitoring by the seismographs over an extended period of time noted that this background ringing was itself slowly becoming less and it was this that gave scientists a clue to the date of the supposed event that set it going in the first place. Assuming the vibration to have decayed at a fairly constant rate over time, they estimated that the Moon was set ringing sometime around the latter part of the twelfth century. In other words, about the same time as the Canterbury monks saw their flaming torch erupt from the side of the Moon!

A fresh impact crater was discovered near the site of the "torch" as well as indications that a large meteorite impact had actually occurred around the same time. Did the monks really witness a giant meteorite or small asteroid slamming into the Moon?

Many astronomers think so. For instance, Dr. Duncan Steele has even attempted to identify the meteor complex to which this object may have belonged. Noting that the date of the event is close to that of the Beta Taurid meteor stream, which is itself associated with the short-period Comet Encke and (possibly) several asteroids, Steele suggested that the Moon was probably struck by a fragment that long ago broke away from that comet. He also argued that the devastating meteorite that blew up over the Tunguska region of Siberia in 1908 was probably another fragment (but more about this elsewhere in this book) and, somewhat ironically in view of the role lunar seismographs played in this saga, he includes a cluster of impacts registered by these instruments in June 1975 within the complex as well. He argued that these events give evidence of a dense swarm of meteorites, including some disturbingly large ones, within the more diffuse Taurid complex. As a chilling final thought he remarked that, although he does not pretend to know what year the world will end, he would be willing to bet that the month it ends will be June!

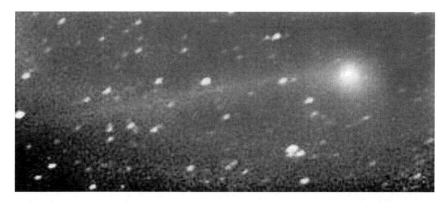

Parent of the Canterbury Swarm! Comet Encke during its 1997 return. © Gordon Garradd 1997.

In honor of the monks who witnessed the 1178 event, this hypothetical cloud of flying rocks and mountains is now known as the "Canterbury Swarm."

Not all astronomers are convinced, however.

Well known British astronomer, popularizer of astronomy, and lunar expert Sir Patrick Moore is one who strongly doubted that the Giordano Bruno crater was formed as recently as 1178. It is certainly a young crater, but "young" in the context of the lunar surface – innocent as it is of the erosion that constantly assails the surface of Earth – could still be millions of years old.

Probably the biggest objection to a giant lunar impact occurring in historical times (besides the extreme rarity of such events) is the problem of what happened to all the material that an impact like this would fling out from the Moon. Countless millions of small fragments (maybe not all *that* small in many instances!) would have been blown out into surrounding space. Some would have crashed back onto the Moon and some should still be orbiting the Sun along paths quite similar to that of the Earth/Moon system itself. We might expect to be running into the odd one of these even today, but the real show would have come shortly after the lunar impact occurred. Earth should then have experienced the father and mother of all meteor storms, a phenomenal display having an intensity not unlike the great Leonid events that grace our skies from time to time. Indeed, in a paper to *Meteoritics & Planetary Science* in 2001, Paul Withers of the Lunar and

Planetary Laboratory referred to previous calculations implying that some ten million tons of lunar ejecta should have arrived in Earth's atmosphere during the week following the alleged impact. From this, he calculated that Earth should have experienced a week long meteor storm of comparable intensity to the peak of the 1966 Leonids! This 1966 storm peaked at around 40 meteors per second over a time interval of 20 min. Something that kept this up for a full week could hardly have passed unnoticed!

Moreover, whereas the Leonid meteors are only breadcrumb-like particles from a comet, the hypothetical storm following a giant lunar impact would be comprised of real rocks. Some of them might be worryingly large. We might expect Earth to have been swept by something not unlike the fictional "meteor shower" that blinded humanity in *The Day of the Trifids*, except that even that storm only lasted for a single night. Many of the projectiles would probably have been large enough to survive passage through the atmosphere and land as meteorites. Some of these may even have been large enough to blast out craters.

Needless to say, nothing remotely resembling such an event was recorded during1178, and it is safe to say that nothing of the sort happened. The absence of a deluge of rocks from outer space in the twelfth century is a serious objection to the suggestion that a large object struck the Moon at that time.

Well, then, if the Canterbury monks did not witness an impact on the Moon, what *did* they see?

The most likely possibility is simply a large exploding meteor within our own atmosphere, that just happened to be passing in front of the Moon – as seen from the perspective of Canterbury – when it blew apart. If it brightened very quickly just prior to the ter-minal burst, it may not have drawn the monks' attention until just before it reached the limb of the Moon, at which point it exploded spectacularly and appeared like an eruption of flame from the Moon itself. Furthermore, the darkening of the Moon is probably better explained in this view. It resulted from the expanding, thick, dust train left after the meteor exploded. The reported "writhing" of the crescent Moon is not easily explained in terms of something hap-pening at the distance of the Moon itself, but it might describe the satellite's appearance as seen through a turbulent tube of gas within Earth's atmosphere.

Although similarities should not be pressed too far, the disturbance of the Moon's image as reported by the Canterbury monks reads a little like something seen by Harden Schaeffer in Florida on the night of June 6, 1973. Harden was observing Jupiter through his 8-in. (20-cm) telescope at 400 magnification, when a jet plane just happened to pass in front of it!

Not the best thing to happen, but what followed was nevertheless very interesting.

About a second later, Jupiter began to shimmer and develop a "jagged appearance." This disturbance settled down after 4 or 5 s, but then a smooth straight line "looking like a crease in a picture" slowly passed over the planet in a direction perpendicular to the flight of the aircraft itself.

Of course, this air turbulence occurred at a lower altitude than that of a meteor, and was on a different scale, but the distortion of the planet's image as observed by Mr. Schaeffer and that of the crescent Moon as seen from Canterbury all those years ago sounds at least superficially similar.

Something "local" like a meteor also accounts for the otherwise curious omission of the event in all other records from that year. Although possible, it does seem strange that an event as spectacular as a giant meteorite or small asteroid hitting the Moon – complete with all the described pyrotechnics – should have only been seen by a single group of five or six people. Is it not likely that somebody else, somewhere, would have noticed something strange about the Moon that evening and recorded it for posterity?

Weired Lights, Mists, Eruptions … and Lightning!

Nothing as spectacular as the 1178 event has since been associated with our Moon, but that is not to say that hosts of other equally strange happenings have not been reported right down to the present day. In many instances, these have been seen and described by experienced observers.

For instance, on the night of April 19, 1787, the famous astronomer William Herschel (discoverer of the planet Uranus) noticed three red glowing spots on the dark part of the Moon. Herschel was convinced that he was witnessing a trio of erupting volcanoes, and estimated the intensity of the brightest of the three as being greater than that of a comet then visible. The comet in question was the one discovered by Pierre-Francois Mechain nine days earlier and which (judging from the rather meager extant descriptions) was apparently very near the limit of naked-eye visibility. From this, we may assume that the lunar bright spot shone with the intensity of a dim naked-eye star. Clearly, this was not a marginal observation that could be brushed aside as an optical illusion.

We now know that genuine volcanism does not occur on the Moon, so Herschel's quite natural interpretation of what he saw could not have been correct. That, of course, in no way casts doubt on the validity of the observations themselves.

It might be worth pointing out that these observations were made during a time of high solar activity. On the very night that Herschel saw the lights on the Moon, auroral activity was noted as far south as Padua in Italy, a long way from its Arctic Circle home. Of course, this may have been pure coincidence, but is it also possible that the particles, ejected from the Sun, that gave rise to Earth's aurora also triggered some sort of gas release from the lunar surface and set it glowing like a mini-aurora?

Whether there is any truth in this speculation or not (and more will be said later about this possibility), Herschel was certainly not the last person to see red glowing spots on the surface of the Moon. From time to time all manner of subtle changes have been reported on the face of our nearest celestial neighbor.

For instance, the noted astronomer W. H. Pickering, who made many observations of the Moon from the hills of Jamaica through the years 1919–1924, noted dark patches that appeared to change shape over a period of time, as well as other patches within the crater Eratosthenes that seemed to move slowly across the crater floor. Never one to shy away from unconventional explanations, Pickering suggested that these observations were evidence of lunar life; the stationary spots that changed shape, he identified with areas of vegetation and the migrating spots with dense swarms of lunar insects or animals! Pickering was quick to point out that he did not believe that life as we know it on Earth could survive on the Moon, and probably not on Mars (regarding the latter, he was being more cautious than some of his contemporaries).

He was clearly not thinking of swarms of lunar locusts or flocks of selenic sheep, nor did he picture massive herds of bison stampeding across vast lunar prairies. But he did speculate that his observations hinted at the existence of something on the Moon's surface to which the term "life" would not be entirely inappropriate.

In addition to these indications of "life" on the Moon, Pickering also claimed to have, like Herschel, witnessed volcanic eruptions and even spouting geysers.

Moving spots were apparently seen by other observers beside Pickering, although some, at least, of these were fast moving and looked more like objects passing between Earth and the Moon than something actually traversing the Moon's surface. These reports go back many years. As long ago as October 15, 1789, Johann Schroter saw something crossing the Moon, and similar reports were made in 1864 and 1873. On April 24, 1874, Safarik was observing the Moon in daylight when he noted the appearance of a bright star-like object visible against the lunar disc. The bright object moved in an ESE to WNW direction, eventually leaving the disc altogether and appearing in the sky "like Vega or Sirius" when seen in daylight. This is very unlikely (to say the least!) to have been related to the Moon itself and was most probably just a piece of thistledown or an airborne spiderling drifting high in the air and catching sunlight.

But back to Pickering! Although he was not alone in reporting such events, very few astronomers found his interpretations convincing.

In fact, many doubted that any changes at all occurred on the lunar surface. At least, nothing more remarkable than the constant play of shifting shadows, probably with some optical illusion thrown in for good measure.

Maybe some of the phenomena were (like beauty is supposed to be) in the eye of the beholder; literally so in the thought of some people. "Floaters" have at times been implicated as being responsible for some of the reports.

What exactly are "floaters"?

This is a name given to dead cells from the retina that literally float around in the vitreous fluid of the eye and show themselves as out-of-focus ghostly spots within the field of vision. Most people have them, and most of the time we take little notice of them, but when looking at a bright background, it is surprising just how many floaters become visible. Needless to say, when viewed through a telescope, the surface of the Moon excels at showing up floaters, and the idea was floated (sorry!) that these might have been mistaken for moving objects on the Moon itself or, more likely, as things passing between Earth and Moon. At best, however, this suggestion could account for only a very small percentage of reports.

Yet, immune to all skepticism, a trickle of reports of odd lunar phenomena continued, some of them from very reputable astronomers. Thus, Walter H. Haas observed a "milky luminosity" on the wall of the crater Tycho in the late 1940s, and on a February night in 1949, F. H. Thornton reported seeing "a puff of whitish vapor obscuring details for some miles."

Even more mysterious was the strange account given by N. J. Giddings, of the Bureau of Plant Industry Soils and Agricultural Engineering, (Riverside, California) concerning a strange phenomenon he witnessed on the early evening of June 17, 1931. In Giddings' own words,

> I was working in the yard near our house at Riverside, California, and happened to glance at the Moon. It was an unusually fine, clearly outlined new Moon, and as I stood looking at it, suddenly some flashes of light streaked across the dark surface, but definitely within the limits of the Moon's outline. Since this was a phenomenon which I had never seen before, I continued to watch it and saw similar flashes streak across the Moon again in a moment or two. Without mentioning what I had seen, I called my wife's attention to the new Moon. She admired it. When I asked her to watch it closely to see if she noticed anything strange, she said "Oh, yes, I see lightning on the Moon," adding that this appeared to be confined to the Moon. We watched it for some 20 or 30 min during which the phenomenon must have occurred at least six or seven times.

Mr. Giddings added that he wrote to Mount Wilson Observatory regarding the phenomenon, but "[their] reply very courteously discounted my observations."

It is difficult to find an explanation for this strange report. Meteorite impacts on the Moon have been suggested. Professor Duncan Steele, for example, noted the date in June and wondered if this may not have been yet another incidence of meteorites associated with the Taurid complex striking the Moon. Impacting meteorites would, however, only be visible as points of light rather than as lightning-like flashes. Moreover, if there were that many large meteors striking the Moon at the time, why were their companions missing Earth?

By the way, even if they cannot explain the Giddings incident, meteorites hitting the Moon certainly cause brief flares, and it is possible that just such an event was photographed in 1953 by Dr. Leon Stuart through an 8-in. (20-cm) reflector. More recently, flares made by meteorites striking the Moon were imaged during the Leonid meteor storm of 1999, and yet others have been imaged since that time, including the impact of what was probably a rather large Taurid on November 7, 2005. This latter object was estimated to have been about 4.7 in. (12 cm) in diameter and to have blasted out a small crater nearly 10 ft across and about 15 in. deep. These lunar events – i.e., bright point flashes of very short duration – no longer present a mystery.

PROJECT 1
Lunar Meteorites

Although the chances of seeing a meteorite strike the Moon are small, they are not zero, and we should always keep this in mind when observing our satellite through a telescope.

The best time to watch is during an active meteor shower, such as the Perseids in August or the Taurids in November. The latter is not a rich shower, but because it has its share of large meteoroids, it is capable of producing some rather hefty impacts.

Flares caused by impacting meteorites are, not surprisingly, more easily visible on the Earth lit portion of the Moon or on the darkened portion during a partial or total lunar eclipse.

Any suspected flare should be timed as accurately as possible and the position on the Moon correlated with other simultaneous reports.

The Lunar Section of the Association of Lunar and Planetary Observers is always interested in hearing from anyone who suspects a lunar impact. If two or more people run a coordinated program, so much the better! ALPO rates as a confirmed observation an impact observed by at least two people separated by at least

30 miles (50 km), within 2° of latitude and longitude on the Moon and within 2 s of time. A tentatively confirmed observation is one by two or more observers separated by less than 30 miles, within 5° of longitude on the Moon and within 5 s of time. An event is deemed probable if captured on video by a single observer, provided it appears in two or more frames and has a star-like appearance. A single observation by one observer, but with a confidence of 50% or greater, is simply known as a "candidate."

Even so, things are not always as they seem. What appeared to be the flash of a meteorite striking the Moon, and reported as such by George Kolovos and colleagues in 1988, was later shown by Paul Maley to have been a flash from the artificial satellite DM SP F3. The satellite just happened to be transiting the Moon when its panels caught the sunlight!

Meteors passing through Earth's atmosphere and transiting the Moon were also raised as a possible explanation of the Giddings event. Writing in *Science* in 1946, James Bartlett pointed out that at the time of the Giddings' sighting, the Sun would have only been set for around 25 min and the western sky still very bright. Bartlett suggested that what the Giddings saw was a parade of meteors passing in front of the Moon. He argued that the background sky

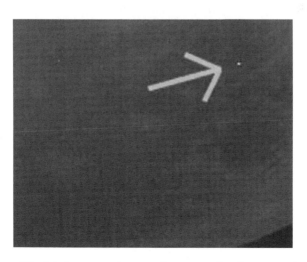

All Over in a Flash! A meteorite crashes into the lunar surface May 20, 2006. NASA image.

would have been too bright for them to be seen except during their brief flight across the Earth-lit part of the Moon.

The problem here is the frequency of their appearance. To see one meteor after another crossing the same Moon-sized patch of sky "six or seven times" within "20 or 30 min" would require a meteor storm dropping an awful load of meteors! Assume that six events occurred every 20 min. This translates to a rate of 18 *events* (i.e., 18 times that a meteor transited the face of the Moon) per hour. As the Moon is approximately half a degree in diameter, this implies that a circle of sky of 1° should have been crossed by a meteor about 36 times each hour. Making a *very* rough estimate of the average length of the meteors' paths as 10°, each of these meteors would (on average) have been occupying 10 square degrees of sky. Taking 2,000 square degrees as the field of view for a single observer, anyone in a suitably dark sky at that time should have been seeing around $(2,000/10) \times 36 = 7,200$ meteors per hour – at least as good as the spectacular Leonid storms of 1999 and 2001! A moderately strong storm such as this would surely have been noted in other parts of the world where the skies were darker and even at the Giddings' site as twilight deepened.

We will take note of just one more early report before looking at the event that launched the "modern" period (as we might call it) of anomalous lunar happenings. The report to which we are referring could be called "The Crow in the Moon" and hailed from Mr. Frank B. Harris who saw it on the (Saturday) night of January 27, 1912.

According to Mr. Harris, "About 10.30 [PM] … I was surprised to see the left cusp showing the presence of an intensely black body about 250 miles long and 50 wide, allowing 2,000 miles from tip of cusp to cusp. The appearance was … in shape like a crow poised." Harris noted that the he kept the object in view until 2 a.m. after which time he ceased observing because of the intense cold. Cloudy weather prevailed until the following Tuesday, by which time the black mark had vanished!

Whatever this was, we can be certain that it was *not* a crow!

The Alphonsus "Eruption" and the Pink Cobra's Head

Two events occurring early in the second half of the last century became watersheds in the story of anomalous events on the Moon.

Until then, reports like those we have been looking at were regarded more as curios than solid observational facts. They were "believe it or not" tales that at times may have seemed to possess a ring of truth, yet always managed to elude the grasp of scientific proof. If accepted, they demanded a change in the way people thought about the Moon, and such shifts in our thought patterns are never something to be undertaken unless the evidence is solid. Lunar anomalies never managed to solidify enough!

Then, on November 2, 1958, Russian astronomer Nikolai A. Kozyrev observed what looked like an "eruption" on the central peak of the crater Alphonsus. This event lasted for about half an hour. So far, this sounds like just another observation of an unlikely lunar event by a reliable astronomer, but this time there was an important difference. The 48-in. reflecting telescope being used by Kozyrev was equipped with a spectrometer with which he managed to obtain a series of intriguing spectrograms. These appeared to show the presence of gaseous emissions identified with the spectral bands of the C2 and C3 molecules. While obtaining a second spectrogram, Kozyrev noted "a marked increase in the brightness of the central region and an unusual white color." Then "all of a sudden the brightness started to decrease." This time, the spectrum was normal, i.e., simply sunlight reflected from the lunar surface without any indication of emission lines.

Although some astronomers have since cast doubt upon the reality of the emission bands in the spectrum, the fact that Kozyrev at least appeared to have obtained some harder evidence than mere visual observation seemed to place lunar anomalies on a firmer footing. But explanations did not come readily. A full-fledged volcanic eruption was not likely, although an eruption of a pocket of gas trapped below the Moon's surface looked credible. (But where did the gas pocket come from in the first place?)

Another idea that was eventually mooted was an impact by a tiny comet fragment. This possibility was suggested by the carbon emissions (which are regular parts of comet spectra), but the chance of a mini-comet scoring a bull's-eye on the central peak of Alphonsus seemed too farfetched to be taken seriously. Moreover, Kozyrev claimed to observe further activity within the crater subsequent to the 1958 event as, for instance, on October 23 the following year. Clearly, whatever was happening in Alphonsus was not due to impacting comets!

The Kozyrev incident was, however, just a prelude to an event observed 5 years later by a group of very experienced astronomers. It was this event, more than any other, which helped to legitimize the thorny topic of lunar anomalies.

On the night of October 29, 1963, two Aeronautical Chart and Information Center cartographers, James A. Greenacre and Edward Barr, working at Lowell Observatory, recorded a remarkable phenomenon on the southwestern side of a hill known as Cobra's Head. This hill can be found to the southeast of a lunar valley known as the Vallis Schroteri and the southwest interior rim of the Aristarchus crater. The phenomenon was very colorful, with bright red, orange, and pink specifically mentioned.

Although this was a visual (as distinct from photographic) observation, the reputation of Greenacre was such that most astronomers apparently accepted his word without question. According to Willy Lea, "The first reaction in professional circles was, naturally, surprise, and hard on the heels of the surprise there followed an apologetic attitude, the apologies being directed at a long-dead great astronomer, Sir William Herschel." In the words of Winifred Sawtell Cameron, "This ... started the modern interest in observing the Moon."

No sooner had the shock of the October event began to settle than another remarkable observation was made. This second took place at the Pic-du-Midi Observatory in the French Pyrenees on November 1 and 2 and involved the expert lunar observers Zdenek Kopal and Thomas Rackham. These astronomers photographed "a wide area of lunar luminescence," and Kopal's *Scientific American* article discussing the event resulted in this becoming one of the most widely publicized of all anomalous lunar episodes.

The events of October and November 1963 finally placed the subject in the mainstream of lunar research and made a large contribution to the revival of interest in our natural satellite over the coming years. Coincidentally, the year of these happenings was also the year that U.S. President John F. Kennedy committed his country to landing a man on the Moon before that decade's end.

The astronomical community's shift of attitude toward these lunar phenomena is exemplified by the changing opinion of British

Moon expert Patrick Moore. An ardent observer of our natural satellite since his youth, Moore was for a long time understandably skeptical about reports involving colored spots or patches. Not having seen anything of the sort during his many hours of Moon observing, he had his doubts about their reality, although keeping an open mind as to their possibility. In the 1961 edition of his well-known book *The Amateur Astronomer*, Moore noted that "blues, greens, and reds" were reported on the Moon from time to time, but added that most of these reports came from "observers using small refractors." This caveat is significant, as small refractors are not the most reliable instruments when it comes to observing color, thanks to a phenomenon known as chromatic aberration. Briefly, this is a false color effect arising from the difference in the degree of refraction of light having different wavelengths. On passing through a glass lens, red light refracts less than blue because of its longer wavelength. Because of this, red light is brought into a focus a little further from the lens than light of shorter wavelengths. Yellow light will be a little closer to the lens and blue will be closer still. Of course, the differences do not amount to much, but they are sufficient to turn what we had hoped to be a white-light image into a collection of overlapping images of different colors. In reality, each image becomes a little rainbow.

Now, a bright star seen through a simple refracting telescope may look pretty with all its colors separated, but the accuracy of the observation is badly compromised. For this reason, all astronomical refracting telescopes worthy of their salt use achromatic lenses, i.e., lenses comprised of several layers of glass having different densities and, therefore, different refractive indices. In theory, these lenses recombine the wavelengths back into white light but, although a remarkable improvement is made, it is difficult to get rid of the chromatic aberration problem altogether and astronomers generally do not place as much faith in color estimates made with refractors as they do in reports using reflecting telescopes, where chromatic aberration is not an issue. Moore's mention of small refractors in association with reports of colors on the Moon hints at where he thought the explanation may have lain.

He did, however, mention the Kozyrev observation of 1958 and the subsequent reports of October 23, 1959, though remarking that on the latter occasion he was also observing the area through an 8.5-in. (21-cm) reflector without seeing anything unusual. On the other hand, he apparently was convinced that Kozyrev did see something of interest in 1958 and suggested that "the serious amateur may carry out useful work by keeping a close watch on Alphonsus to see if any further disturbances occur there." His wording seems to hint that the activity within this crater (if real) was of a localized nature rather than an example of a more wide-spread phenomenon.

Moore's only other reference to anything remotely approaching transient events was a brief remark about alterations of dark patches within large dark-floored craters such as Grimaldi. He noted that there had been reports of these changing shape, spreading and changing tint during the course of a lunar day. He also mentioned that "tiny craterlets inside Plato exhibit unpredictable fluctuations, being sometimes conspicuous and sometimes completely invisible." He did not elaborate on these changes, however, nor speculate as to why they might have occurred.

Yet, within 7 years of the publication of this edition of *The Amateur Astronomer* Moore, together with other lunar experts Barbara Middlehurst, Jaylee Burley, and Barbara Welther compiled a *Chronological Catalog of Reported Lunar Events*. In the meantime, he had not only witnessed a transient red coloration himself (presumably while using his *reflecting* telescope!) but also became credited with coining the term by which these events are now known – transient lunar phenomena, or TLPs.

How can TLPs be explained?

No one knows for sure, and it is quite possible that no single explanation will fit all cases. But that has not deterred a variety of ideas from being aired.

The early supposition that TLPs are volcanic eruptions can almost certainly be ruled out. The Moon is simply not the place for active volcanism.

Similarly, explanations involving objects impacting on the lunar surface cannot account for the majority of the reports, although, as we have already seen, it is certainly the favored explanation for brief point-like flashes.

Another suggestion involves the release of pockets of gas from beneath the Moon's surface. Excitation by sunlight, according to this hypothesis, causes the released cloud of gas to glow briefly like the contents of a neon tube, before it disperses into

Moon map showing locations of recorded transient events. NASA image.

space. This sounds plausible, although a credible origin for such gas pockets is not immediately obvious.

On the other hand, the skeptical attitude of the early critics has been maintained by some astronomers, who see nothing more substantial than optical illusion at work here. This position, however, is not easily maintained in the face of the sort of evidence presented earlier.

PROJECT 2
Transient Lunar Phenomena

Because these are such fleeting events, the chance of seeing one while casually looking at the Moon through a telescope is not high, although it is always possible. Careful scrutiny of the regions of the Moon that have shown most activity in the past gives the individual observer the best chance of catching one of these fleeting episodes, but it is even better to have several people watching designated areas as often as possible in the hope that more than one person will observe the same thing from different localities.

Large telescope size is not essential, and any good quality reflector of 6″ (15 cm) or more should be quite sufficient.

If an event is observed, determine its position on the Moon's surface as accurately as possible and note the time of the occurrence, also as accurately as possible, and the nature of the event (i.e., red glow, apparent mist, or whatever). If two or more people report the same thing at the same time and in the same place, it is very hard to dismiss it as illusory!

If you happen to see something, the Lunar Section of ALPO would be very interested in the details. But even if you do not, simply finding the lunar locations where these events have been reported is an interesting exercise in its own right!

One explanation that may hold promise for at least certain classes of TLPs involves "dust storms" created by the electrostatic suspension of dust particles above the lunar surface. Actually, there is some independent evidence that this process actually occurs, although its association with TLPs is less than firmly established.

While in orbit around the Moon, astronauts on board *Apollo 8, 10, 12,* and *17* reported seeing "bands" and "twilight rays," a little like the familiar crepuscular rays seen after sunset and before sunrise on Earth. Like terrestrial crepuscular rays, these appeared just prior to lunar sunrise and just after lunar sunset. The Surveyor spacecraft actually photographed "horizon glows" that seem to have been of a similar nature.

On Earth, crepuscular rays are caused by shadows of clouds or mountains below the horizon of the observer. The light sky between the bands of shadow and the bands themselves radiate outward from the position at which the Sun has set or from which it is about to rise. Of course, for such a pattern to be visible, the Sun's light must pass through some medium. In the case of Earth, this is the atmosphere and its constant haze of suspended particles of one sort or another. But, as the Moon has no atmosphere to speak of, how can such a phenomenon be visible there?

The answer appears to be "dust haze."

Among the instruments left on the Moon's surface by the *Apollo* astronauts was a device designed to monitor dust particles kicked up by impacting meteorites. Known as the Lunar Ejecta and Meteorites, or LEM, it was left behind by the *Apollo 17* crew in 1972.

Remarkably, the LEM found more than had been expected. Each (lunar) morning relatively large numbers of particles were detected moving from east to west at speeds lower than what would be expected for meteorite ejecta.

What could be causing this movement of dust?

In the opinion of Timothy Stubbs of the Solar System Exploration Division at NASA's Goddard Space Flight Center, the culprit is probably electricity. He suggests that the day side of the Moon is positively charged and the night side negatively. At the interface between day and night – the lunar terminator – "electrostatically charged dust would be pushed across the terminator sideways" by horizontal electric fields.

If this is correct, the terminator must be forever accompanied by a long and narrow electrostatic dust storm!

Although this does not obviously explain TLPs, one may speculate that some regions of the Moon are coated with greater amounts of fine dust than others and that the terminator dust storm might become especially dense in these places. Could this be responsible for reports of "obscuring mists" that come to light from time to time? May there not also be breakaway "clouds" of dust that could sometimes catch the sunlight at certain angles and briefly appear bright against the lunar surface?

Electrostatically raised dust is unlikely to explain all transient Moon mysteries. But it may at least account for some of them.

"Protuberances", Bridges, and Other Lunar Oddities

Other strange things have been reported on the Moon from time to time which, although not TLPs in the strict sense, nevertheless deserve mentioning.

Of these oddities, probably the oddest would have to be the "protuberances" seen in July 1875 by A. J. Loftus and E. C. Davidson from on board the *Coronation* in what was then known as the

Gulf of Siam but is more commonly known these days as the Gulf of Thailand.

On July 13, Loftus and Davidson noticed "a prominent projection ... with the naked eye on the Moon's upper limb." This was apparently confirmed when "the best glasses on board were ... brought to bear upon it." The witnesses noted that "The protuberance, in color, was similar to that of the Moon." The Moon was about 20° altitude at the time.

The protuberance had disappeared by the following night; however, Loftus noted that a smaller one had by then appeared at a different region of the Moon's limb. "This," he continued, "had also disappeared before the Moon rose on the evening of the 15th." when the Moon "finally presented its usual unbroken appearance."

Curiously, a report by an anonymous correspondent to *Scientific American* on January 28, 1882, apparently referred to a similar appearance seen the previous year. Coincidentally(?) this also happened in July (on the third of that month) and was said to have been witnessed by several residents of Lebanon, Connecticut.

The Moon was nearly full and, according to the correspondent, about an hour high when the observers noticed "[T]wo pyramidal luminous protuberances ... on the Moon's upper limb. They were not large, but gave the Moon a look strikingly like that of a horned owl or the head of an English bull terrier." They were said to have been a little darker than the rest of the Moon's face. Unlike the ones reported in 1875, these protuberances "slowly faded away a few moments after their first appearance, the one on the right and southeasterly quarter disappearing first."

This was not the end of the matter, however, as "About three minutes after their disappearance two black triangular notches were seen on the edge of the lower half of the Moon. These points gradually moved toward each other along the Moon's edge, and seemed to be cutting off or obliterating nearly a quarter of its surface, until they finally met, when the Moon's face instantly assumed its normal appearance. When the notches were nearing each other the part of the Moon seen between them was in the form of a dove's tail."

There can be little doubt that this latter report had more to do with Earth's atmosphere than with the Moon per se. Regarding the earlier report, although the appearance was certainly longer lasting, it is difficult to see how anything other than our planet's atmosphere was responsible for that appearance as well. It may be significant that, in the first report, the Moon was said to have been about 20° above the horizon at the time of the first sighting and, in the second instance, about an hour (i.e. around 15°) high.

Earth's atmosphere was not, however, the culprit behind another lunar mystery, one which unfortunately developed some distinctly odd interpretations and associations over the years.

We refer to the so-called "moonbridge," first noted by John O'Neill (then science editor of the *New York Herald Tribune*) on July 29, 1953. O'Neill believed that the feature he saw through his telescope was a sort of natural rock bridge, not unlike those in the American West, except for its prodigious length of several kilometers. The feature is located at (lunar) latitude +17°, longitude +50°, and best manifests when the Moon is about three days past the full.

The exact circumstances of O'Neill's observation are difficult to reconstruct, but the bridge he believed he saw seemed to have spanned two formations then known as Promontorium Olivium and Promontorium Lavinium on the shore of Mare Crisium. The two "Promontorium" epithets are no longer used, and there seems to be some discrepancy about precisely which features were so named, but the location of the alleged bridge itself is clear enough.

Unfortunately, the bridge proved to be rather shy in showing itself to other observers, although famed Moon expert H. P. Wilkins did apparently see it, greatly strengthening the case for its reality.

The main observational support for the bridge was a fan-shaped area of sunlight apparently emanating from a low point between P. Olivium and P. Lavinium, strongly giving the impression of being caused by the Sun shining under a huge natural arch of rock. This seemed quite a straightforward explanation. After all, if something waddles and quacks, chances are that it is really a duck, and if something looks like a bridge of rock with sunlight streaming under the arch, chances are it is exactly that. Moreover, in the early 1950s, there did not seem to be any reason to doubt that such a thing could exist on our nearest neighbor.

Yet, not everyone was happy with this explanation. Even Virgilio Brenna, whose 1963 book *The Moon* contained a spectacular artist's depiction of the bridge as supposedly seen from the lunar surface, had his doubts about the feature's reality. He suggested that it might be nothing more than an illusion created by the interplay of light and shadow.

In fact, by the time Brenna wrote this, the "illusion" explanation was pretty well established. As early as January 1954, Paul Rocques of Griffith Observatory used a 12-in. (30-cm.) refractor to photograph the region of the bridge and, through analysis of these photographs, concluded that the fan-shaped patch of sunlight which O'Neill and Wilkins had interpreted as being caused by the low Sun shining under an arch, resulted instead from "sunlight coming through a pass and over the sloping shoulders of the promontories, falling on rising land westward."

Other observers tried for the feature, but it appears that few if any actually saw anything, and a few were tactless enough to suggest that Wilkins was getting too old to see properly! That sort of remark was cruel. Wilkins is remembered, and rightly so, as the greatest selenographer of the pre-*Apollo* era, and the hypothesis of a natural rock arch was actually a very reasonable one. It may have been wrong, but it was still a good hypothesis. Alas, following denigrating remarks from some members of the British Astronomical Association, Wilkins felt that he had to resign from that association with which he had for so long shared his expertise.

With the advent of the space age, our knowledge of the Moon has exploded, and we now know that Rocques' interpretation was correct. There is no moonbridge. Unfortunately, though, that has not stopped the subject from having become absorbed into UFO literature of the more crackpot variety. Of course, it has been reinterpreted in a way that would have angered O'Neill and Wilkins. Sensationalist writers tried to turn the bridge into proof of intelligent life, something that neither O'Neill nor Wilkins ever contemplated.

The at times rather sad saga of the moonbridge stands as a warning, not only to the fact that even experts can misinterpret observations and make honest mistakes, but also that others can turn these honest and (if it may be so expressed) "conservative" mistakes into sensationalist drivel with the potential for bringing the whole issue into disrepute.

PROJECT 3
Seeking an Illusion *

Looking for something that isn't there may seem a waste of time, but readers with relatively large telescopes and some experience at lunar observing may like to see if they can duplicate the O'Neill/ Wilkins illusion.

The following images should enable the site of the alleged "bridge" to be located, but a telescope of at least 16" (41-cm) is probably about the minimum size required. The best time for looking is said to be three days after full Moon.

Is there anything visible at or near this spot that looks like sunlight shining under a natural rock arch? If you see anything suspicious, it might be of interest to see whether different filters enhance or diminish the illusion.

Remember, if you see anything that could be interpreted as a bridge, you have joined a very elite club of observers. As far as we can tell, the "club" still has a membership of two!

Unidentified Floating Objects: Bodies Seen Transiting the Moon

Earlier, mention was made of some old reports of objects seen moving across the face of the Moon and, on one occasion, remaining visible against the sky outside the satellite's limb. These events probably represent nothing more mysterious than debris raised to relatively high altitudes in our atmosphere by updrafts, but occasionally something is reported crossing the Moon that makes us wonder if wafting debris and thistledown (or the notorious "floaters" in the eye) really tell the whole story.

A case in point was reported by W. Steavenson at the Royal Greenwich Observatory in 1920. At the time, Steavenson was

(a) The Full Moon, showing the region of the alleged "moonbridge". NASA image. (b) Close-up of the moonbridge region. The "bridge" was thought to span the low point between Pr. Olivium and Pr. Lavinium. In these images, S is at *top* and W to *left*. Lunar Orbiter image.

observing the lunar crater Plato with the observatory's 28-in. (71-cm) equatorial telescope when he noticed "a small black object [entering] the field on the North side and [passing] nearly centrally across it in an upward direction." He estimated the time taken for the object to cross the field (itself 6 min of arc across) to be about 2 or 3 s. The object itself appeared as a dot about 1 s of arc in diameter and was in perfect focus.

Steavenson did not, at first, pay much heed to the object, admitting that he had previously seen apparent "transits" of dark objects projected against the Moon and had always dismissed them as either specks of dust drifting across the diaphragm of the eyepiece or as distant birds. However, as he later mulled over this particular event, he started doubting whether either of these explanations really did it justice.

Writing in the *Journal* of the British Astronomical Association, he noted that "The eyepiece was a positive one, and was focused on the wires of the micrometer, of which it gave, of course, an erect image. It follows, therefore, that the object, if it were a speck of dust, must be moving *upwards* in the plane of the wires, which seems highly improbable... On the other hand, one cannot exclude altogether the possibility of internal [air] currents in such a large tube."

"But if," Steavenson continues, "the object were not in the plane of the wires, it must have been outside the telescope altogether. In that case it is possible to get some idea of... its minimum distance. The focus of the objective is 28 ft and the telescope was focused on the Moon... for an object 10 miles away the eyepiece would require a shift of about 1/6 of an inch [to bring it into focus]."

Steavenson notes that a shift of even half this distance would be enough to put the Moon's image right out of focus, yet he was adamant that both object *and* Moon were in perfect focus. This, he maintained, must mean that the object (if it really was outside the telescope) was at least 20 miles (32 km) away. This makes the bird explanation unlikely.

He next raises the possibility that the body may have been a meteor. However, considering its relatively slow motion across the face of the Moon, a meteor would need to be traveling almost head on toward the observer, which would be an enormous coincidence to say the least. Furthermore, at a minimum distance of 20 miles, the meteoroid would need to be at least 6 in. (15 cm) in diameter to be seen as an arc-second-sized dot against the bright background of the Moon. However, an object of that size at the supposed minimum distance would not be dark. It would be visible as a brilliant fireball. If further away, it would need to be still larger and would create an even larger fireball upon entering the lower atmosphere. Yet no fireball was reported during the night in question.

Interestingly, the following year something very similar was seen by Mr. R. Moran while observing the Moon through a 6-in. (15-cm) reflecting telescope at a magnification of 40 times. Moran sent his report to *Popular Astronomy* and gave no indication of having any knowledge of the Steavenson observation.

In Moran's own words, "Whilst viewing the Moon… a small black dot appeared on the disc of the Moon and traveled across the Moon's disc in about 6 or 8 s. The diameter of the black object was probably about 3 or 4 s of arc."

On many previous occasions, Moran had witnessed distant birds crossing the face of the Moon. These were always faster and quite easily identified as birds, and he was quite convinced that the object he saw in 1921 had a different explanation. Just what that explanation might be, though, was a different matter!

Observations such as these have been put forward as evidence for small natural satellites of Earth. From the study of the orbits of Earth-approaching asteroids, we know that some of these can temporally become satellites of our planet; however, most go into temporary Earth orbit well beyond the Moon. Nevertheless, the existence of small natural bodies in Earth orbit between our planet and the Moon is not impossible. In the distant past, we know that giant meteorites struck Earth, the Moon, and neighboring planets with such force that rocks were ejected into space. Rocks from both the Moon and Mars have been found on Earth, and nobody doubts that the exchange was a two-way street. Presumably, some of these rocks blasted from Earth could have ended up in orbits smaller than that of the Moon and might be close enough and large enough to be observed in transit. The trouble is, there is no other evidence that any satellite rocks really are orbiting our planet.

We *do* know however, that large numbers of small Earth-approaching asteroids exist and that some of these can come very close indeed. It might seem a big coincidence that two very small "asteroids" (if we can call something measured in centimeters an asteroid!) should have almost grazed Earth only months apart in the early 1920s, but who is to say that it couldn't happen?

Is that what Steavenson and Moran saw?

An interesting thought, but one that does not seem too outlandish in the absence of more information.

The First Weather Satellite

Earth's first weather satellite was Tiros 1, launched in 1960, right?

Well, in one sense only. This was the first *artificial* weather satellite, but the first object in Earth orbit capable of telling us something about our weather was launched over 4 billion years ago when a Mars-sized proto-planet gave our young world a mighty glancing blow and sent much of its crust hurtling into space. In other words, the first weather satellite was none other than the Moon!

Please don't misunderstand. This is not about folk superstitions such as the one that says there will be rain if the crescent is on its side (tipping the water out!). This is not about the Moon *predicting* Earth's weather at all. No! This is serious stuff.

It is not about future weather so much as present atmospheric conditions. The degree of cloudiness in our atmosphere affects the amount of light reflected by Earth. A hypothetical observer on the Moon would see a brighter Earth when our planet is largely covered by cloud. And the more light our planet reflects onto the Moon, the more it gives right back to us in the form of earthshine. Ergo, a cloudier Earth means a brighter "old Moon in the new Moon's arms." So if we are experiencing one of the clear spots on Earth and we see the rest of the crescent Moon brightly lit by earthshine, we can conclude that much of our planet is under cloud that night.

The effect, however, is subtle – far too subtle to be noticed by eye alone. Even an Earth completely clear of cloud (and there is *always* some cloud) makes a good reflector and makes for relatively bright earthshine.

The first person to take a specific interest in the variability of earthshine's intensity in response to conditions on Earth appears to have been D. F. Arago. In a paper on the subject which he read before the Paris Academy of Sciences on August 5, 1833,

Arago proposed that the intensity of the Earth-lit portion of a crescent Moon could be monitored as an indicator of terrestrial cloudiness. At least, this was possible in theory. At the time of his writing, Arago fully realized how difficult such measurements would be to make in practice, but he speculated that when more sensitive means of measurement became available, "we may be able to read in the Moon the record of the average cloudiness of our atmosphere."

Arago's hope of more sensitive instruments eventually materialized in the form of the visual double-image photometer developed by Andre Danjon (1890–1967) who was, incidentally, like Arago before him, a director of the Paris Observatory.

When observing through a double-image photometer, one sees two lunar images juxtaposed in such a way that the dark limb of the first just touches the bright limb of the second. The observer can reduce the intensity of the second (bright) image – for instance, by using a calibrated photometric wedge – until the two adjacent areas appear of equal intensity. In this way, the difference between the bright part of the Moon's image and the darker Earth-lit region can be measured. This method has the advantage of being unaffected by atmospheric extinction, haze, or superimposed light.

Using this device, Danjon found that he could monitor the Earth-lit portion of the Moon's disk until three days after first quarter. In effect, he could determine the difference in brightness between Sun and Earth for the various phases of the latter as seen by a hypothetical Moon-being.

From these observations, he deduced an average absolute magnitude for our planet, i.e., how bright Earth would be if situated one Astronomical Unit from both the Sun and a hypothetical observer. (An Astronomical Unit – AU – is the mean distance of the Earth and Sun, approximately 93 million miles, or 150 million km). The value he derived for Earth's absolute magnitude was −3.92, but he also discovered that there are seasonal variations, implying a range in the planet's albedo or reflectivity from 0.52 in October to 0.32 in July. In other words, Earth is a better reflector in October, bouncing an average of 52% of the incident sunlight that it receives back into space during that month. Presumably, there is more cloud around during October and less in July than at other times of the year.

By the way, before leaving the subject of earthshine, we should mention that this explanation for the "Old Moon in the new Moon's arms" actually marked an important advance in our understanding of Earth's true nature as a planet. As we all know, before Copernicus proposed his heliocentric model of the Solar System, the prevailing cosmology understood Earth to be central with the celestial objects orbiting it. As popularly understood, the alternative model of Copernicus dethroned the position of Earth by moving it out of the center of the universe. Also, as popularly understood, the pre-Copernican geocentric model understood the universe to be a small place, with Earth taking up a significant portion of the entire cosmos.

This is not exactly true. For instance, Ptolemy was quite explicit in his belief that the universe was vast. Referring to the distance of the stars, he opined that, by comparison with these distances, Earth appeared as a geometric point. Technically, a point has position but no magnitude. A line consists of an infinite number of points. Therefore, by saying that Earth appears as a point compared with the distance of the stars, Ptolemy is effectively saying that the distance of the stars is infinite. Of course, this is not to be taken too literally. Earth is not really a point. But Ptolemy is saying that *for all practical purposes*, the distance of the stars is infinite. Earth, in Ptolemy's model, may be at the center of the universe, but the universe itself is still immeasurably vast and Earth merely a dust mote at its core.

Moreover, this dust mote was not considered, by the pre-Copernicans, as anything very attractive. Following Aristotle, they thought of it as an agglomeration of "gross" matter that settled into the center of the (otherwise pure and unblemished) universe. Rather than being the apex of the universe, Earth was more like the cosmic garbage heap!

Far from demoting Earth's position in the cosmos, Copernicus actually elevated it. From being the place where all the muck of the universe settled, he now numbered it among the celestial orbs.

From this, it follows that for a hypothetical being on one of the other planets, Earth itself would appear as a "wandering star" or planet. An inhabitant of Mars would see Earth as a planet in the Martian sky just as we see Mars as a planet in our own sky.

But what would an inhabitant of the Moon see?

From the Moon, our planet would be a truly glorious sight. It would appear as a large orb of such brilliance as to illuminate the lunar night. This was how Galileo reasoned. Once Earth became recognized as part of the astronomical universe and not simply a pile of gross matter at its lowest level, the issue of Earth's appearance from other vantage points in the universe became a valid question. If it really is a planet, then it should be a brilliant object from the Moon and we should be capable of seeing something of its illumination of the lunar night.

And we do!

The faint illumination of the "Old Moon in the New Moon's arms," therefore, became for Galileo evidence that we really are a planet. He argued that this phenomenon is, indeed, evidence that Earth shines brightly on the lunar surface; evidence that we are, as Copernicus's model implies, truly denizens of the celestial spheres.

The Moon acts as a "weather" satellite in another, albeit related, way as well. As anyone with a passion for observing lunar eclipses knows, these can vary greatly in the eclipsed Moon's degree of visibility. Normally, after the Moon has passed fully into the darkest cone, or *umbra*, of Earth's shadow, it remains quite clearly visible as a copper-colored disk. The amount of light received from the eclipsed Moon is normally not enough to perceptibly brighten the night sky, but if it was concentrated into a point rather than spread out over the half-degree lunar disk, it would still look pretty bright.

Yet, not all eclipses are equal. Records show that during an eclipse in 1848, the Moon remained so bright that many people refused to believe that an eclipse was even in progress! Writing in the *Monthly Notices of the Royal Astronomical Society*, one Mr. Forrester noted that "during the whole of the... lunar eclipse of March 19, the shaded surface presented a luminosity quite unusual, probably about three times the intensity of the mean illumination of an eclipsed lunar disc." He also noted that there was a brilliant aurora at the time, and wondered if that may have been responsible for the brightness of the eclipse.

At the other end of the scale, the Moon was said to have disappeared altogether during the eclipse of 1761. Something similar happened in 1963. The writer recalls watching the eclipse of December

30 that year through a 2.5-in. (0.6-cm) refractor and being quite unable to see the eclipsed region of the Moon at all. The edge of the shadow, as I remember it, had a distinct greenish appearance. During totality, it was just as if there were no Moon in the sky.

PROJECT 4
Estimating the Brightness of a Lunar Eclipse

The degree of brightness of a lunar eclipse is measured on a 0–4 point scale known as the Danjon scale. The darkest eclipses are given a value of 0, the brightest of 4.

The Danjon scale is given in Appendix A. Estimates are made by comparing the appearance of the eclipsed Moon with the descriptions on the scale.

Good quality estimates are always welcomed by the Association of Lunar and Planetary Observers, as well as by interested climatologists such as Dr. Keen (see Project 5). With nothing more than the naked eye, it is still possible to make scientifically useful observations!

These variations tell us much about the transparency of Earth's atmosphere at the time of the eclipse. The air must have been especially clear in 1848, but something thick was obviously about in 1761. An unusually dark eclipse in 1950 was blamed on the amount of smoke in the atmosphere from extensive forest fires in Canada, and the culprit for the very dark 1963 one was the major volcanic eruption on the island of Bali earlier that year. The extensive pall of stratospheric dust from this event was also responsible for some of the most spectacular sunsets that this author has ever seen.

Clearly, the brightness (or darkness!) of the Moon during a total lunar eclipse provides a good diagnosis for the haziness of Earth's atmosphere at the time. Although there may not be any studies on this, perhaps it is safe to say that there has been a general trend of brighter eclipses in recent decades as the worst of the former industrial smoke pollution has been improved. Apparently, the decrease in particulate pollution (mostly smoke from factory chimneys) over Europe has been sufficiently marked in recent years to contribute slightly to higher temperatures over much of the Northern Hemisphere. Ironically, environmental awareness has contributed to global warming!

Of course, variations in atmospheric opacity caused by other events (an unusually cloudy season, volcanic eruptions, large forest fires and dust storms) must be factored in, but it would be interesting to see if a general trend monitoring the declining level of industrial pollution is detectable.

PROJECT 5
Estimating the Stellar Magnitude of the Eclipsed Moon *

During recent years, in a project being run by Dr. Richard Keen, a number of amateur astronomers have been carefully monitoring the brightness of the Moon during eclipses as a means of determining the atmosphere's turbidity. Using nothing more than naked eyes and small pairs of binoculars it is possible to directly measure the average state of our planet's atmosphere! This is an interesting and useful project capable of being carried out from the backyard or even through a bedroom window!

The project goes beyond simply assigning the eclipse a rating on the Danjon scale. It attempts to quantify the eclipsed Moon's total brightness in terms of the stellar magnitude scale.

To estimate the brightness of the eclipsed Moon in terms of stellar magnitudes, it is necessary to shrink the apparent size of the Moon's image down until it looks not too different from that of a star. The best way to do this is to view the Moon through the reverse end of a pair of binoculars and compare it with a star as seen with the naked eye.

However, the two cannot be directly compared. A correction factor must first be taken into account.

Looking through the reverse end of a pair of binoculars shrinks the Moon's size by a factor equal to the magnification of the binocular. For instance, looking through a reverse 10× binocular will reduce its size by 10 as compared with the naked-eye view (just as looking through the correct end will *increase* its size by 10) and will reduce its apparent brightness by a factor of 100, i.e., by five magnitudes. So, to compare the Moon's shrunken image with a naked-eye star, the corrected value would be the "raw" brightness as compared to a star (or stars) of known brightness, "brightened" by five magnitudes. As the brighter stellar magnitudes have the smaller numerical values, this means that the true brightness of the eclipsed Moon is the "raw" value minus five magnitudes.

For example, suppose the Moon's shrunken image looks as bright as a star of first magnitude as seen by eye alone. The real or corrected magnitude of the eclipsed Moon would then be

$$1 - 5 = -4.$$

The real brightness of the eclipsed Moon would be –4, about equal to that of Venus.

Readers with a background of variable star observation are especially encouraged to participate in this program. Seriously interested readers are asked to contact Dr. Keen at Richard. keen@colorado.edu. All observations should include the time of the observation and details of the size and magnification of the binoculars employed.

Lunar Eclipse Oddities

The above section has inevitably led us into talking about eclipses of the Moon (see Appendix B for a full list of lunar eclipses from 2011 until 2050). We can see that not all eclipses are created equal. Some are dark and some quite light. Now, we will take a look at a few that also seem to have been a bit weird!

What, for instance, are we to make of the odd shape apparently assumed by Earth's shadow as seen by a certain "H. H." writing in *Nature* on August 18, 1887?

According to this correspondent, during the lunar eclipse of August 3 that year, the eclipsed portion of the Moon appeared "flattened" along its leading edge. Observing from Hamburg, "H. H." noted that a small cumulus cloud appeared below the Moon, and at first he assumed that the darkened segment was part of this. However, an hour later, the cloud had disappeared, but the eclipsed

Partial phase of lunar eclipse. © Easy Stock Photos.

region of the Moon maintained its flattened appearance. This, he said, was observed by "several persons" other than himself.

Another correspondent to *Nature*, replying to the first letter and calling himself simply "M. C.," also noted that the appearance of Earth's shadow was peculiar during the August 3 eclipse; not so much flattened as "irregular and jagged." This correspondent remarked that "the appearance was certainly unusual; at least I never saw anything like it."

On the other hand, a third correspondent, H. P. Malet, wrote that "from Killin, on Loch Tay, the shadow on the Moon had no form similar to that given by 'H. H.' and suggested that the alleged peculiarities reported were due to nothing more exotic than clouds. Maybe, but the persistence of the effect would tend to count against this as well as the insistence of both 'H. H.' and 'M. C.' that the sky was clear (excepting the transitory cumulus cloud early in the first correspondent's observation)". The fact that the eclipse appeared "normal" from Killin does hint strongly, however, that the effect was "local" in some way, either a trick of the atmosphere or some perceptual oddity ... although who knows what this may have been!

A different oddity was observed during the eclipse of July 6, 1963, by Captain T. H. Davies of the *Canopic*, while en route from Sydney to Aden. By the way, the eclipse referred to here was a partial one and is not to be confused with the very dark total eclipse of December 30 mentioned above.

Evidently, this earlier eclipse was not as dark as the December one, as Davies mentions that "At maximum eclipse when three-quarters of the Moon was in shadow, its surface still remained visible." What was strange however was the appearance of "fingers of light ... illuminating the upper section which was in shadow."

At first glance, this observation may seem better placed among TLPs; however, the cause was apparently quite different. Commenting upon the observation, H. B. Ridley of the British Astronomical Association wrote that "The Earth's atmosphere refracts light from the Sun into the shadow cone, so that a lunar eclipse ... is not complete; the Moon is still quite plainly visible even when wholly immersed in the Earth's shadow. The atmosphere scatters blue light ... but transmits red: therefore the faintly visible Moon has a coppery hue ... The 'continental' areas of the Moon are much brighter than the flatter, darker maria or 'seas,' and show up very plainly during eclipse".

"It is fairly evident that what the observer saw ... was the comparatively bright north polar region, partly illuminated even though in the Earth's shadow".

"Although there is nothing exceptional about this observation, the officer concerned was quite justified in remarking on the phenomenon, which might have escaped the notice of a more casual observer."

Another eclipse of unusual aspect and well observed from sea was that of January 29–30, 1953. The unusual feature of this event was a system of multiple colored bands crossing the face of the eclipsed Moon and well observed from ships at sea. Some 18 ships reported seeing the eclipse, of which six saw the full total phase. All six of these reported the unusual color display. It seems that, progressing from top to bottom of the Moon's disk, there were bands of color ranging from faint white through light blue through to green and light yellow to light orange. Of these the bands of blue and green were considered the most unusual, as these are colors that one would not normally expect to be refracted by Earth's atmosphere.

Nevertheless, reading of the reference to green certainly caught the present writer's eye. Recall the very dark 1963 eclipse and the mention that the advancing edge of the shadow gave the impression of a distinct greenish hue? This may have been a contrast effect from using a refracting telescope at the time which, though color corrected, may not have been entirely reliable when objects as bright as the Moon were in view. Still, when all is said and done, the edge of the shadow *did* look green (well, sort of green)!

The Moon: Our Lifesaver?

From the above accounts, we see that our good old Moon is not the uninteresting orb sometimes depicted in elementary astronomy texts. Close though it may be by the standard of astronomical distances, it continues to retain its share of mysteries.

In fact, increasing numbers of scientists are coming to believe that it plays a vital role in the greatest mystery of all; that of our home planet's habitability. Not just a provider of romantic moonlit nights, the Moon's presence may be essential to our very existence! So, before leaving our satellite for more distant cosmic fields, we should take a look at this somewhat weird connection between the Moon and our lives here on Earth.

Earth is very fortunate to have the Moon. According to the most widely accepted hypothesis, the presence of a large moon (or of any moon, for that matter) orbiting our world was an unlikely occurrence. Our glorious Queen of the Night is, according to the best evidence, the by-product of an unlikely grazing collision between proto-Earth and a Mars-sized wandering planet in the days of the Solar System's youth. Had the interloper's trajectory been just a little steeper, the collision would not have been "grazing" and the infant Earth would probably have been smashed to smithereens.

On the other hand, had its approach been a little shallower, it would have missed Earth altogether. Either way, the evolution of our planet would have been very, very different. In the first instance, Earth may have become an asteroid belt or (if the pieces of both planets had managed to come together once more) a moonless rocky world quite different from the one we know and love today.

In the second instance, Earth would have avoided catastrophic impact but would have been forever moonless.

Birthing the Moon. Born of catastrophe! Two proto-planets collide and the Moon is formed. © William K. Hartmann, March 2003.

Resent research suggests that a moonless Earth may be a far less congenial alternative than we might imagine. In 2004, British geologist Dave Waltham found evidence that Earth is very finely balanced between enormous changes in its axial tilt accompanied, on one side of the scale, by a dramatic increase in its period of rotation and, on the other, by an equally dramatic spin-up of its rotational velocity. Without going into the intricate details (anyone desirous of studying these is directed to *Astrobiology* 4, pages 460–468, where Waltham's paper "Anthropic Selection of the Moon's Mass" is published), Waltham concludes that it is the Moon's presence that keeps our planet poised on this narrow balance. A straying to one side or the other, while probably not rendering Earth sterile of life per se, would likely make it uninhabitable for human beings. According to his calculations, an increase of a mere 2% in the Moon's mass would result in axial wobbles as great as 50° over periods of just a few millions of years ("short" on geological timescales). Even an increase of just 1% would be enough to cause its rotational speed to dramatically slow and the days to drastically lengthen. Conversely, a *decrease* in the Moon's mass of similar magnitude would have the opposite effect; a sharp increase in Earth's rotational speed and a shortening of days.

In the first instance, the climate would experience extreme fluctuations over relatively short periods of geological time, courtesy of the planet's shifting rotation axis. As for the effects of slower rotation, average wind speeds would decrease, night/day temperature differences would increase, while the temperature gradient between equator and poles would become much less pronounced. Melting of the polar icecaps would appear to be just one consequence of this.

Conversely, speeding up the planet's spin makes for stronger winds, less diurnal temperature range, and a steeper gradient between equator and pole. The latter may have had an especially chilling (ah!) consequence during the time of the so-called "snowball Earth" period just prior to the emergence of the first multicellular organisms on this planet. At that time, polar temperatures came close to the freezing point of carbon dioxide. On a faster spinning Earth, with consequently lower polar temperatures, this strong greenhouse gas may have frozen out of the atmosphere and built up as a cap of dry ice covering the familiar water-ice ones that we know today. Earth would then have mimicked Mars in that respect and the depletion of CO_2 from our atmosphere may have prevented the eventual thawing from the "snowball" epoch. Earth, to this very day, may then have been one vast ice sheet of short frigid nights and days, devoid of any life more complex than algae!

Such a fine balance, by the way, neatly explains an otherwise improbable "coincidence" that is sometimes remarked upon in astronomy books and then passed over without further comment. We refer to the odd match between the apparent angular diameters of the Sun and Moon as seen from Earth's surface.

This essentially exact match becomes very apparent during total solar eclipses. A significantly greater disparity of angular diameters of the two players in these celestial dramas would mean that total eclipses would either fail to occur (and phenomena such as prominences and the solar corona would remain invisible) or would be so "deep" that these phenomena risked being completely covered by the Moon's disk.

It is not too strange to think of the Sun's apparent size as being fixed. A larger/nearer sun or a smaller/more distant one would mean differences in the amount of solar heat received by our planet, and it increasingly appears that only a slight increase or decrease in this would be sufficient to trigger a catastrophic runaway greenhouse effect on one hand or an equally catastrophic runaway glaciation on the other. Either way, we would probably not be around to "enjoy" a smaller or larger Sun.

But if Waltham is correct, it seems that the angular size of the Moon as we see it in our skies is equally fixed by similar "anthropic" considerations. The spectacle of a total solar eclipse, complete with visible prominences and the glorious corona, is an added bonus ... and a most welcome one at that!

2. Odd but Interesting Events Near the Sun

Of the major planets in the Solar System, only two can pass across the face of the Sun, at least as seen from the vantage point of Earth. Such a phenomenon is called a "transit," and the only planets that can accomplish this feat are the two so-called "inferior" planets, Mercury and Venus.

By the way, there is nothing discriminatory intended by calling these planets "inferior." All it means is that they are located between Earth and the Sun, "beneath" Earth's orbit with respect to the Sun.

Transit Tales: Regular and Weird

Venus

Transits of Mercury are the more common, but those of Venus are the more spectacular and have generated far greater scientific interest, especially during the earlier years of the Scientific Revolution.

It was the famous British astronomer Edmond Halley who first suggested, in 1716, that transits of Venus could be used as a means of more accurately determining Earth's distance from the Sun. This distance is known as the Astronomical Unit and is the main yardstick for the moderate distances found within the Solar System.

Back in 1619, Kepler worked out the *relative* distances of the planets from the Sun. Since that time, it was known that (for instance) Venus orbited at a mean distance of 0.72 that of Earth, Mars at 1.52, and so forth. But the *absolute* distances (the actual number of miles or kilometers) was only very roughly determined.

D.A.J. Seargent, *Weird Astronomy*, Astronomers' Universe,
DOI 10.1007/978-1-4419-6424-3_2, © Springer Science+Business Media, LLC 2011

This situation could be changed if some method of measuring Earth's distance from one of the nearer planets could be found. If the true distance between our world and (say) Venus could be determined, by already knowing what the relative distances are, our absolute distance from the Sun could be easily calculated.

Halley saw that the accurate measurements of transits of Venus could achieve this end.

How could observing transits do the trick?

Briefly, because of the phenomenon of parallax, two people viewing the same transit from two relatively widely separated locations on Earth's surface would see the transiting planet take a slightly different path across the face of the Sun. As Venus is relatively close to Earth, this difference is measurable.

Because the Sun is perceived as a disk, this also means that each observer sees the planet's track as differing slightly in length, and it is the accurate measuring of the difference in length that the value of the parallax is determined, and from this, the planet's true distance derived.

To achieve this end, Halley proposed that two widely separated observers accurately determine the time of first contact (the first appearance of the planet's limb against the Sun's surface), second contact (when the planet breaks free of the solar limb), and the third and fourth contact at the end of the event.

The precision of this method is, however, compromised somewhat by a phenomenon known as the "black drop," about which more will be said shortly. This made it all but impossible to obtain the required accuracy in the timing of the planet's "contacts" with the solar limb.

Nevertheless, the pursuit of these rare events proved a good stimulus for expeditions to remote parts of the world. It also inspired great voyages of discovery such as that by Captain James Cook.

Nowadays, the scientific value of transits has largely disappeared. But they are still worth observing simply for the wonder of seeing the black dot of Venus (visible – with adequate protection, of course – to the naked eye) cruising slowly across the face of the Sun.

It is a pity that whole generations pass without having the chance to witness such an event. The transits actually happen in pairs, but with such long gaps between them that the observers of one pair have long departed this life before the next pair comes along. Actually, just seven transits of Venus have been observed since the invention of the telescope: on December 7, 1631, and December 4, 1639, June 6, 1761, and June 3–4, 1769 (Captain Cook's transit), December 9, 1874, and December 6, 1882, and, most recently, June 8, 2004. The next is due on June 5–6, 2012, after which there will be no more until December 10, 2117, and December 8, 2125. Notice that the entire twentieth century passed without seeing a transit of Venus.

At our time in history, transits occur in pairs according to a pattern that repeats itself every 243 years. A pair of transits separated by 8 years occurs, followed by a transit-free gap of 121.5 years, then a second event just 8 years later followed by a gap of 105.5 years.

As mentioned, the black silhouette of Venus against the brilliant face of the Sun can easily be seen by the naked eye. Of course, the eye should never be completely "naked" when looking toward the Sun, and unless suitable protective measures are taken when viewing such an event, the penalty of carelessness will be blindness.

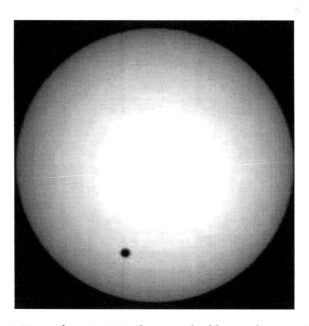

Venus transit December 6, 1882 photographed by students at Vassar College.

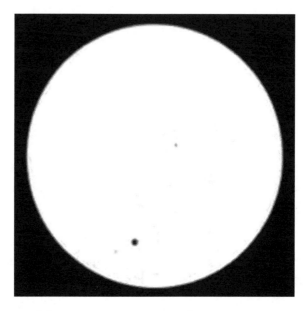

Another image of the Venus transit December 6, 1882.

The disk of Venus near the solar limb June 8, 2004. © Gordon Garradd 2004.

Through a telescope (also adequately protected, of course) the event can be eerily spectacular. An unusual effect, visible especially when a low-power eyepiece is used for the observation, is the apparent tendency for the silhouetted planet to remain connected to the Sun's limb even after its entire disk has progressed onto the face of the Sun. This is the so-called "black drop," which proved such a menace to the exact timings required by those trying to refine the measurement of Earth's distance from the Sun. This phenomenon, by the way, owes more to the atmosphere of Earth than to that of Venus, as was originally thought. The effect has also been noted (though, by the nature of things, in a less conspicuous form) during transits of that smaller inferior planet, Mercury, which has no atmosphere worthy of the title and certainly none capable of giving rise to any phenomena observable during transits.

As we will see below, weird things have been reported from time to time in association with transits of Mercury. However, probably on account of the rarity of Venusian transits, few mysterious sightings have been reported during these events.

Nevertheless, the transit of 1874 did bring forth a couple of odd observations. For whatever reason, both of them came from observers located in New South Wales, Australia.

Thus, we have the observation by Mr. L. A. Vessey, who reportedly saw the planet during transit not as a black silhouette but

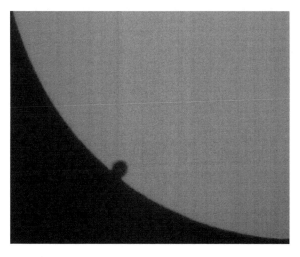

Venus in transit June 8, 2004 showing *"black* drop". NASA image.

as having a distinctly grey appearance, albeit with a black center. Also, during that same transit, Sydney Observatory astronomers H. C. Russell and H. A. Lenehan noted a bright spot of light near one of the planet's poles as it crossed the face of the Sun.

These two anomalies were probably not associated. The first may have been merely a contrast effect, but the bright spot is unlikely to be explicable in that way. It is difficult to say what it may have been, although if it was very near the dark rim of the planet, it might have been a high-altitude Venusian cloud forward scattering the Sun's light. At least, that *sounds* plausible!

They Missed the Transit but Found the Bird!

Arguably, some of the most interesting scientific results of Venus transits were those having nothing to do with astronomy at all! We have already mentioned Captain Cook and his voyages of discovery, but these were not the only serendipitous advances courtesy of Venus transits.

For example, as the transit of 1761 approached, the renowned French naturalist Charles Buffon found himself, in his capacity as head of the Acadamie des Sciences, in charge of sending astronomers and their equipment to distant observing sites. One of the more hazardous expeditions was reserved for the noted astronomer-monk, Alexandre-Gui Pingre, and his assistant Denis Thullier, to the remote island of Rodrigues; a 24-mile-long pile of basaltic rock in the middle of the Indian Ocean.

Buffon himself may have been a little envious of Pingre and Thullier, not because of the transit and certainly not because of the sea voyage to the ends of Earth, but (as a naturalist) he was intrigued about the flora and fauna of the strange place to which they were bound. So, just in case the transit was clouded out, he instructed the two astronomers on a contingency plan, providing instructions for a thorough collection of plant and animal specimens and the recording of all that appeared unique to the island.

Just as well, as it turned out!

Come Transit Day … and the sky filled with clouds. Realizing that observing the transit was a lost cause but anxious that the expedition would not be a complete waste of time and effort, Pingre put Buffon's "contingency plan" into action and began the biological survey of Rodrigues. The results of the survey, although never published as a book, are preserved in the Bibliotheque Sainte-Genevieve in Paris.

Included in this work is Pingre's account of a large flightless bird known, thanks to its seemingly unsociable nature, as the Solitaire (*Pesophaps solitaire*). This bird was actually a relative of the Dodo and, alas, has now gone the way of its famous cousin. Pingre apparently did not actually see the Solitaire with his own eyes, but he was assured by an inhabitant of the island that a small number still existed there at the time. This second-hand report is our only evidence that the bird existed on Rodrigues as late as 1761.

At the time, this was little more than a curiosity. Detailed knowledge of this avian oddity did not come until 1874; courtesy of another transit of Venus!

Gone the way of the Dodo. Alas! Illustration of the now-extinct Solitaire *Pesophaps Solitair.* From "Extinct Birds" 1907, Courtesy Wikipedia.

This time, an expedition of British scientists chose the island as a favorable spot to observe the transit and, just like their French predecessors, their interests extended beyond the astronomical. Sadly, by that time there were no living Solitaires – not even stories of living Solitaires – but the scientists did find bones. Solitaire bones aplenty. In all, 15 complete skeletons of the bird were finally assembled and the information gleaned from these enabled, for the very first time, the Solitaire to be established as a separate species of bird, something that had previously been the subject of debate. And all because of two transits of Venus!

But what of this second transit? Was it observed, and did it add to astronomical as well as to ornithological science?

It was observed, but scientifically it was disappointing. Little material of astronomical value was brought back by the Rodrigues expedition!

PROJECT 6
Observing the Transit of 2012

On June 6, 2012, the second and final transit of Venus in the present century will take place. Over a century will pass before the next one, so if you miss it, there will not be another chance to see one in your lifetime!

This map shows the region of Earth from which the transit will be visible. The Roman numerals denote first (I), second (II), third (III), and fourth (IV) contacts. First contact is when the disk of the planet reaches the Sun's limb, second when the planet's disk breaks free of the limb and its silhouette is fully on the solar disk, third contact is when the leading edge of the planet's disk touches the limb of the Sun as the silhouette begins to leave the disk, and fourth contact is the final exit from the solar disk.

The transit will be visible with the naked eye, although (it must be repeated) the eye must NEVER be completely "naked" when viewing the Sun. If you intend trying for it without optical aid, only use proper solar filters, *NEVER* smoked glass or even welding goggles. If directly viewed through a small telescope, even greater caution needs to be taken. Only recommended filters, *PLACED OVER THE OBJECT GLASS AND VERY SECURELY STRAPPED ON*, should be used.

2012 Transit of Venus

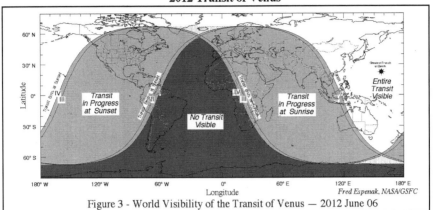

Figure 3 - World Visibility of the Transit of Venus — 2012 June 06

Map showing region of visibility of the Venus transit of June 6, 2012. NASA image.

The best method for watching the event is by projecting the Sun's image through a small telescope onto a white card. It is best to use a small refractor rather than a reflector. In general, reflectors are not suitable for any kind of solar work unless the mirror is left unsilvered, which of course makes it useless for anything else. Even refractors should be small, or they will focus too much heat. A 2 or 2.4 in. (50–60 mm) object glass is quite large enough, and anything over 4 in. (100 mm) or thereabouts should be stopped down to 2–2.4 in. An opaque cardboard cap over the object glass with a circular hole cut to the required diameter will be sufficient for this.

It is fascinating to watch the large black dot drift slowly across the face of the Sun. Watch out especially for the Black Drop effect and, of course, for the possibility of anything unexpected.

If you enjoy watching the Venus transit, don't despair that no more will occur in your lifetime. Although not as spectacular, remember that a transit of Mercury is due in 2016!

Mercury

These are less spectacular than transits of Venus and have never been held in the scientific esteem of the latter. Nevertheless, they can at least be watched with the knowledge that many observers of earlier events are still alive and well. I have seen a few myself, missed some others, and am still well under one century old!

The next transit of Mercury takes place in 2016. A full list of the remaining twenty-first century transits is given in Appendix D.

Being a smaller and more distant planet, Mercury's silhouette against the solar disk is much smaller than that of Venus and cannot be seen with the naked eye. A small telescope will suffice, however, and the safest way watching the progress of the tiny black speck is by projecting the image of the Sun onto a white sheet of paper. For my most recent observation of a transit of this planet, I found that the back of a letter from the local government authorities made an excellent screen for projecting an image from a 2.5-in. (6-cm) refracting telescope!

PROJECT 7
Observing the Mercury Transit of 2016

For the first time in nearly 10 years, Mercury will transit the Sun on May 9, 2016. The event will be visible from the Americas, Europe, Africa, and central Asia, and first contact is scheduled to occur at 14:57 h Universal Time.

Unlike Venus, Mercury is not visible with the naked eye when in transit, but may be seen as a very small black point through large binoculars such as 15×80 s. These must, however, be properly mounted, and the safest and easiest way of watching the transit with these instruments is by projecting the solar image onto a sheet of white paper.

For a larger view, a 2.5-in. (0.6-cm) refracting telescope (once again, properly mounted) at a magnification of 50–100 should be sufficient. Because the Sun is being observed, too much aperture means too much heat, so this is not a job for large telescopes.

At first contact, the limb of the planet first touches that of the Sun, and shortly afterward it becomes visible as a small notch on the solar limb.

Watch for any "black drop" that may occur as the trailing edge of Mercury breaks free of the solar limb at second contact. If you are viewing the event directly through a safe solar filter (CAUTION: Make sure that it really is a *safe* solar filter!), keep watch for any effect (real or illusory) similar to the bright spots and so forth occasionally reported at earlier transits. The chances are slim, but there is always a chance that something interesting might be seen, even if it is just an interesting optical illusion!

Unlike transits of Venus, the relative frequency and low scientific interest of Mercury's passages across the Sun's face has not inspired scientific expeditions to distant lands where bones of extinct birds might be found. These Mercurian events lack the romance of the Venus transits!

Nevertheless, odd things have been reported from time to time. In fact (undoubtedly because of the greater frequency of these events), transits of Mercury have managed to chalk up more anomalous observations than those of Venus!

Thus, during the transits of 1799, 1832, 1848, 1861, and 1868, a bright spot on the planet was noted by such eminent astronomers as Schroter, Harding, Kohler, and Professor Moll.

But what is to be said for haloes and rings seen surrounding the planet during transit?

Although a trace of atmosphere does exist, it is hardly worthy of that title and is entirely too rarefied to cause any such phenomena as haloes or rings of light. Nevertheless, just such a halo was reported during the 1707 transit by none other than the Assistant to the Astronomer Royal, and similar sightings were noted during the transits of 1753 and 1786. Then, in 1799, no less an astronomer than Schroter reported a halo "scarcely brighter than the surface of the Sun, but of another color" surrounding Mercury's silhouette. A similar ring was observed from the Royal Observatory during the transit of 1832 and described as having "a violet hue, the color being strongest near the planet." Then again, William Huggins noted a bright aureole – said to have been a little brighter than the Sun's disc, during the transit of 1868.

Curiously, according to a study by B. G. Jenkins published in 1878 (more will be said about this study below), when Mercury transits the Sun in May, dark and nebulous rings are reported, but when it transits in November, bright haloes are seen.

Most probably, the reported haloes were nothing more than simple contrast effects; a suggestion made as long ago as 1878 by an anonymous contributor to the *Monthly Notices of the Royal Astronomical Society*. This contributor had actually witnessed the halo effect himself. Describing his own observations of the transit of May 6 of that year, he noted that the planet appeared surrounded by a halo, much brighter than the surface of the Sun, and having an irregular outline. About the same time, E. Dunkin also observed a halo, albeit fainter than the surface of the Sun, that "appeared with radiating arms as in the solar corona." In agreement with the anonymous observer, Dunkin also explained this as arising from the contrast between the intensely black disc of the planet and the brilliant background of the Sun. Everything that has subsequently been learned about Mercury upholds this explanation. Still, these apparently anomalous observations of a Mercurial halo stand as reminders of how threshold visual observations can easily lead to the postulation of physical effects, which in the end turn out to be spurious.

Are the bright spots, to which we gave passing mention earlier, any more substantial?

In an article published in an earlier issue of the 1878 *Monthly Notices of the RAS*, B. G. Jenkins analyzed the reports of anomalous bright spots observed during earlier transits in the hope of finding patterns capable of predicting events associated with the forthcoming May 6 event. Summarizing his findings, Jenkins concluded that:

- During May transits, when Mercury is furthest from the Sun, a luminous spot appears on the planet's trailing hemisphere.
- The spot is never central but always south of the planet's equator.
- In some transits, two spots occur close together. At others, only a single spot is reported.

Comparing the "well observed phenomena" associated with the May transit of 1832 and those seen at the November transit of 1868, Jenkins found that in the first event a diffuse spot of light preceded the planet's centre but during the second transit, a sharply defined spot followed the center. He also found that a dark ring was reported around the planet in the first instance and a bright halo in the second.

From this, Jenkins concluded that the approaching May 6 transit should show a rather diffuse and ill-defined blob of light (not a point-like bright spot) "gradually sinking from a grayish-white to the dark color of the disc, situated a little in advance of the centre and to the south of it, and ... the planet will be surrounded by a dark nebulous ring, not a bright one."

We have already seen that he struck out on the second prediction. Both Dunkin and the anonymous observer quoted above saw a *bright* halo on May 6, not a dark one!

So how did Jenkins fare with the other prediction?

No better it would seem!

The same anonymous observer who saw a halo brighter than the surface of the Sun (albeit explained by him as a contrast effect) also noted a "minute bright [spot]" very near the center of Mercury's disc. He went on to explain that the spot was "slightly diffused, but with a brilliant star-like nucleus" and, although almost centrally placed, was very slightly on the *trailing* side of the planet.

In other words, Jenkins' predictions were about as far from the mark as possible. If we wish to be very charitable, we might concede that he did at least get the diffuse appearance partially right, but this was more than undone by the far more conspicuous appearance of a bright point at the core of the "bright spot."

So what are we to conclude from all of this?

Before we even try to reach any firm conclusion, it is well to remember that Mercury is a very small planet and that even when at its closest to Earth, it remains a miniscule dot in modest telescopes. When seen projected against the brilliant surface of the Sun during a transit, it is never more than a very small black spot. Trying to discern details on such an object under the conditions of a solar transit is, putting it mildly, not an easy task.

Moreover, it is worth noting that the evidence for bright spots on the transiting planet seems always to have come from early visual observations.

It is very difficult to understand how a bright spot could appear near the center of the transiting disk of Mercury. Accordingly, it seems that, like the halo reported to have sometimes surrounded the planet during transits, this phenomenon owes more to optical illusion than to any physical cause on the planet itself. Maybe someday a bright spot will be recorded by modern imaging, but in the meantime it seems that (in common with so many early anomalous details reportedly seen on other planets) the Mercurial spots must be placed behind the eyes of the beholder rather than located on the surface of another world.

Mercury progresses across the face of the Sun (Image 1). Note the conspicuous sunspot visible beneath the transiting planet.

Mercury progresses across the face of the Sun (Image 2).

Mercury progresses across the face of the Sun. Note the conspicuous sunspot visible beneath the transiting planet.

Mercury progresses across the face of the Sun. Note the conspicuous sunspot visible beneath the transiting planet. Images courtesy John Walker.

Some Transit Trivia

Before leaving the subject of transits, here are some pieces of trivia that the reader might find interesting. Remember them the next time a star party is clouded out!

When a planet passes in front of the Sun, the latter is dimmed by a tiny amount. A transit of Venus causes the Sun's light to drop by 0.001 magnitudes, one of Mercury by just 0.00003 magnitudes. Both amounts are far too small to be noticed.

See how many guests at your next star party know that one! Or this one!

How often do grazing transits of Mercury occur? (A grazing transit is one where some regions of the world see only a partial transit of the planet, that is to say, it never becomes completely clear of the Sun's limb. All the while, the full transit is visible from other regions of Earth).

The answer is … not very often. The most recent occurred on November 15, 1999, but the previous one was as long ago as October 28, 1743. The next will be on May 11, 2391. Readers may wish to mark that on their calendars.

It is also possible for a transit to be only partial as seen from Earth, the full transit in these events missing our planet altogether. The most recent was on May 11, 1937 and the previous on October 21, 1342. Another will not happen until May 13, 2608.

Grazing transits of Venus also occur but (surprise! surprise!) only very rarely. One took place (according to calculations; nobody saw it) on December 6, 1631, but the next is not due until December 13, 2611.

Partial transits of Venus happen from time to time as well. There was one, so the mathematicians inform us, on November 19, 541 BC, and there will be another on December 14, 2854.

Now for something *really* rare. Simultaneous transits of both Mercury and Venus!

These do happen, but don't hold your breath waiting. A double-bill is predicted to occur on July 26 in the year 69,163 and another in 224,508. Although the two planets will not transit together, mark the date of September 13, 13,425 in your diaries for a pair of transits by both planets just 16 h apart.

The date of July 5, 6757 might also be worth noting, as an eclipse of the Sun will occur while Mercury is in transit. The event is predicted to be visible from eastern Siberia. This will also be a double transit of sorts – of Mercury and the Moon.

We have a longer wait for a transit of Venus to coincide with a solar eclipse. The next event of this type is predicted for April 5 in the year 15,232. There was, however, *almost* a Venus transit/solar eclipse double in 1769. The day after the transit, on June 3, a total eclipse of the Sun was visible from North America, Europe, and the northern parts of Asia.

It is possible for a pairing of both Mercury and Venus transits to coincide with a solar eclipse, there is no known prediction as to when this will happen. Maybe the world won't last that long!

At the beginning of this chapter, we said that only two major Solar System objects can transit the Sun from the perspective of Earth. There are, however, minor objects capable of this feat.

In recent years, increasing numbers of asteroids passing within Earth's orbit have been discovered, and some of these must transit the Sun from time to time. All of these objects, however, are very small, and it is not surprising that no asteroid transit has been observed.

Likewise, many comets pass within the orbit of our planet, although most of these visit the inner Solar System only infrequently. Nevertheless, several were computed to have transited the Sun during the past 100 years, although most of these were not discovered until they emerged into the evening or morning twilight after the event had already taken place. Another (the Great March Comet of 1843), although seen before the transit, was not observed sufficiently well for an orbit to be calculated until after the event. It was only then that a transit was found to have happened. In only two instances (the Great September Comet of 1882 and Halley's in 1910) were attempts actually made to observe a transit, and in neither instance was anything seen against the face of the Sun. Significantly, the solid nucleus of Halley is rather large by cometary standards and was passing quite close to Earth at the time of the transit in 1910.

So, besides the rare transits of Mercury and those very rare ones of Venus, Earth's inhabitants are extremely unlikely to see anything passing across the face of the Sun – birds, airplanes, and balloons excepted!

But sometimes in the history of astronomy, the unexpected happens – or (at least) *seems* to happen. People *have* reported objects other than Mercury and Venus transiting the Sun! At one time, astronomers felt so sure that an intra-Mercurial planet existed that they even named it and listed it within the catalog of Solar System objects. Let's have a look now at the tale of the planet that never was.

The Little World that Wasn't There

How does that rhyme go?

> Late last night upon the stair
> I met a man who wasn't there
> He wasn't there again today
> O how I wish he'd go away!

Well, no one knows about the man who wasn't there, but the Solar System was once thought to have harbored a planet that wasn't there. In fact, there was a time when astronomers were so convinced that it *was* there, they gave it a name, calculated its orbit, and included it in descriptive surveys of the Sun's retinue!

Its name was Vulcan, and it was thought to orbit between Mercury and the Sun. Remaining so close to the Sun in the sky, the small planet was very difficult to see under normal circumstances, which explained why it had never been noticed in the morning and evening twilight. Nevertheless, it was considered potentially observable in transit across the disk of the Sun as well as during total solar eclipses under ideal conditions.

Today the very name "Vulcan" has a certain mythic (indeed, almost mystical) connotation as the home planet of *Star Trek's* Mr. Spock, but the Vulcan spoken about here is a very different world. So close to the Sun, it was expected to be hot enough to melt lead. Not even the indefatigable Mr. Spock could survive on *this* Vulcan!

But why should such a world be imagined in the first place?

Actually, in the middle years of the nineteenth century, there seemed to be extremely good reasons for believing in its existence, and it was this that raised speculation to the point of virtual certainty in the minds of many astronomers.

The chief reason was the orbit of Mercury or, to be more precise, the orbit's persistent refusal to conform to the predictions of Newtonian gravity. Nothing is more annoying to astronomers than a planet whose orbit seems to defy the laws of motion. Uranus had been such a planet, and it was the desire to bring it to heel that led to the discovery of the even more remote world that we now know as Neptune.

Mercury behaved in a manner not unlike Uranus, and it was both natural and logical to conclude that if the anomaly in the latter's orbit was due to the perturbing influence of another planet, the behavior of the former was most probably amenable to a similar explanation.

The problem of Mercury's orbit was tackled in 1859 by Urbain Jean Joseph Leverrier, who applied the same method by which Neptune had been mathematically tracked down. It appeared to work, and on September 12 of that year Leverrier announced to the French Academy of Sciences that a new planet had been mathematically "discovered" in an orbit between Mercury and the Sun.

All that remained was for reliable observations of this new world to be made by astronomers. Incidentally, so confident were the astronomers of the day in the validity of Leverrier's solution that they did not wait for confirming observations to give the planet a name. "Vulcan" – from the Roman god of fire and metal forging – seemed a very good one for what must be a planet of infernal heat.

Remarkably, apparent confirmation was not long in coming. Just three months after Leverrier's announcement, news arrived that a certain Dr. Lescarbault – a physician practicing at Orgeres – had actually observed the transit of a planet-like object across the face of the Sun as far back as March 26, prior to Leverrier's presentation of his mathematical solution. Unaware of Leverrier's work, the doctor had kept his sighting to himself in the hope that he could find an independent, confirming observation before announcing his apparent discovery to the world.

Although no further observations were unearthed, publication of Leverrier's results seemed to provide confirmation of sorts, and Lescarbault finally made his sighting public.

Strangely, Leverrier welcomed the doctor's sighting with less than open arms. He actually took it upon himself to pay Lescarbault a visit, announcing with seeming lack of candor that he had come to determine "either that you have been dishonest or deceived."

What the doctor thought of this bombastic astronomer remains unrecorded (maybe just as well!), but it is clear that Leverrier was not exactly overawed by Lescarbault. Upon requesting to see his astronomical instruments, Leverrier was shown a "chronometer" consisting of a huge pocket watch with only hour and minute hands, a seconds-pendulum consisting of an ivory ball attached to a silk thread hanging from a nail in the wall, and a note pad consisting not of paper (of which the doctor was chronically short) but of a plank of wood on which all calculations were performed and, where necessary, erased with a wood plane. Leverrier was less than impressed!

Nevertheless, despite clearly getting off on the wrong foot, Leverrier slowly changed his mind about the doctor's observation and even used the results he obtained with his admittedly primitive instruments to calculate a rough orbit for the new body. He found that Vulcan orbited a mere 13 million miles (21 million km) from the Sun and had a "year" of just 19.75 days.

Nothing further was seen of Vulcan until March 20, 1862, when an English amateur astronomer by the name of Lummis noted a moving and very planet-like black freckle on the face of the Sun. He is said to have watched the supposed transit for 20 min before being called away by some unspecified (but presumably very important!) "official duties." These observations by Lummis enabled the French mathematicians Valz and Radau to compute an orbit which, encouragingly, came out very close to the one determined by Leverrier from Lascarbault's data.

On the face of it, this apparent agreement on an orbit by different mathematicians working, not just independently of each other but also using different sets of data from two independent observers, appeared to provide strong evidence for the reality of the planet.

Alas, such was not the case. It later transpired that at the time Lescarbault was watching the earlier transit, the French astronomer

Emmanuel Liais was also observing the Sun from Brazil, but saw nothing unusual.

The second transit – that reported by Lummis – ended up faring no better. According to Professor C. H. F. Peters, the black dot that Lummis saw was none other than a small and very round sunspot observed at the same time by himself in America and also by Sporer in Europe.

A minute ago, we said that nothing was seen of Vulcan between Lescarbault's alleged transit observations of 1859 and that of Lummis in 1862. Actually, that is not quite correct, as there was a minor sighting on January 29, 1860, although the wider astronomical community knew nothing of it until the publication of a belated note in *Nature* by F. A. R. Russell in 1876 recounting his observation of something reminiscent of a planet crossing the face of the rising Sun 16 years earlier. As Russell recalled, the Sun shone through a fog so thick that it could be observed "as if through dark glass." He described the spot as being similar to Mercury when seen in transit and noted that the event was witnessed by four people, including himself.

Another observer claimed that he watched a transit of Vulcan (date unknown) while squinting through the (unfiltered) eyepiece of a telescope. Think about this for a minute. What would happen to the retina of somebody's eye while squinting through the eyepiece of a telescope pointed toward the Sun? Please, nobody try this one. Only assess the likely accuracy of an observation allegedly made that way!

It should be mentioned that reports of mystery transits did not wait until Leverrier's work on the orbit of Mercury. An anonymous contributor to the *Proceedings of the Royal Astronomical Society* in 1859 sought to bolster Leverrier's claim by listing reports of alleged transits dating back to early 1761, most of them observed by credible and knowledgeable observers. He actually lists eight events, in 1761, 1762 (two that year), 1764, 1798, 1802, 1819, and1820. The event of 1820 (February 12 that year) is interesting in that the observer claimed that the spot not only looked like a transiting planet but also displayed indications of an atmosphere!

Despite their planet-like appearance, we now know that none of these spots could have been a transiting planet. Presumably, they were simply sunspots. Some may indeed have been illusory.

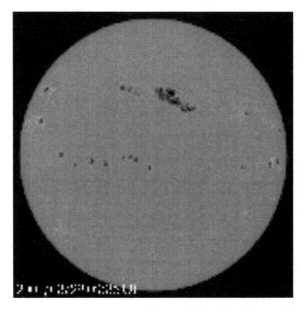

A spotted Sun! Note how the smaller sunspots could be mistaken for the silhouettes of transiting planets. NASA image.

PROJECT 8
Spots on the Sun

Sunspots are easy to observe with only very modest equipment. A pair of tripod-mounted binoculars or small telescope projecting the Sun's image onto a sheet of white paper is all that is required, and on most days at least one spot will be visible.

Monitoring the increase and decrease of sunspots throughout the solar cycle is an interesting project in its own right, but the immediate interest here is to look for spots that may be mistaken for transiting planets. How frequently do spots occur that, in your opinion, might be mistaken for a transiting world? From your own observations of sunspots, do you think that many of the transit-of-Vulcan reports of earlier years might have been influenced by a level of wishful thinking? Have you seen anything that may have fooled *you*?

Nevertheless, not all alleged sightings of Vulcan came in the form of transits. There were also a small number of observations of bright objects seen close to the Sun during solar eclipses. Indeed, observations of this type, if taken at face value, suggested that there might be more than one Vulcan. As we will shortly see, at least one eclipse seemed to produce a pair of Vulcans!

Writing in the prestigious science journal *Nature* in 1878, astronomer J. R. Hind recalled how a group of people at St. Paul's Junction noted two unidentified objects during the total eclipse of August 7, 1869. The best attested object was described as "a little brilliant," a star-like point of light shining through the outer regions of the Sun's corona. This was apparently seen by four members of the group. Attempts to identify this with a known star were made; however, the stars suggested as possible candidates were simply too faint to have been visible under the prevailing conditions.

Even more puzzling were reports by two further members of the group that a crescent-shaped object was also spotted somewhat further from the eclipsed Sun. At least one person claimed that this crescent was observed through a small telescope just before totality and again as the Sun emerged from the total phase of the eclipse. It is not known whether this observer tried to see it during totality, but it is probable that his attention was directed elsewhere then!

A crescent shape is interesting, as an intra-Mercurial planet would go through phases similar to Venus and Mercury itself, but one must also wonder if internal reflection in the telescope (perhaps a ghost reflection of the partially eclipsed Sun itself?) might account for the crescent.

Ultimately though, the mystery of the two objects of 1869 remains unsolved.

The reason why Hind recalled the incident in 1878 is, however, not hard to find. That year also witnessed a total eclipse of the Sun, on July 29, which raised the Vulcan issue in grand measure, as we will shortly see.

Unlike the 1869 incident, the chief players in 1878 were very well known and highly experienced astronomers – no less than Lewis Swift, Professor James Craig Watson, and G. B. Airy, who independently undertook searches of the Sun's immediate vicinity for any signs of intra-Mercurial planets.

They were not to be disappointed, with each astronomer reporting unidentified star-like objects close to the Sun. Indeed, Watson reported finding not one, but two, such objects!

Now, one Vulcan might be acceptable, but *two* of them? Needless to say, this interpretation was hotly debated.

Maybe the rather rushed conditions under which these observations were made, plus the drama and excitement of the event itself, plus the inadequacy of the portable equipment used, all conspired to render the positions of the suspected objects less than accurate. For instance, Professor Peters argued that a small error in the reading of Watson's instruments would bring both his objects into line with two well-known stars.

On the other hand, Swift's object may not be so easily explained.

Like Watson, Swift also saw two objects, and apparently identified one of them with a known star (Theta Cancri). However, the second object corresponded to nothing on the charts. At the time of this observation, Swift knew nothing of Watson's sighting, and when he learned of this the following morning, he saw it as confirming his own results.

According to Swift, the two objects (star and unidentified) were of the same brightness and, from his description, were apparently aligned to the southwest. He described them as red in color.

The sighting by Watson may not actually have been confirmatory, if Peters' corrections (?) were, indeed, *correct*. But another sighting seems to have corfirmed Swift's results!

About the same time, the third independent observer, G. B. Airy, in his own words, "devoted myself to a search for an intra-Mercurial planet." As if to avoid the type of criticism leveled against Watson by Peters, he specifically noted that "In order to expedite the record of position, I placed disks of cardboard on the circles of the equatorial, and marked the pointings by means of a sharp pencil and a pointer. All danger of error from wrong circle-readings is in this way avoided."

The search appeared to pay off, with the discovery of an uncharted object a little further to the east of Theta Cancri, the same star that Swift determined lay close to *his* own mystery object. Like Swift, Airy described the color of his object as reddish but, unlike Swift, noted that it was "very much brighter" than Theta Cancri. He also added the very interesting comment that it "had a perceptible disk" under a magnification of 45 power.

Airy took such great care in determining the position of the object that there hardly seems room for error, especially as the perceptible "disk" would seem to preclude any possible confusion with a star. Although Airy thought it "highly probable" that this was indeed the elusive Vulcan, he did allow that it may actually have been a comet, as "when the tail of a comet and the small appendages of its head are invisible [against the bright sky], the nucleus is usually circular." This alternative possibility now seems the most likely explanation for the object seen by Airy and Swift.

Intriguingly, the position given by Airy is quite close to that of a member of the so-called Marsden Group of comets identified in images from the SOHO spacecraft launched in 1996. These tiny objects are closely associated with another SOHO comet group – the Kracht Group – and several meteor showers. The comet groups are named for Brian Marsden and Reiner Kracht, who first identified them.

The present writer appears to have been the first to note an apparent similarity between the orbits of the Marsden Group comets and those of the June Arietid meteor stream, which in turn is associated with the Southern Delta Aquarid meteor shower of July and also, possibly, the Quadrantid meteors of January. All of these have been linked together with the short-period comet 96P/Machholz (or Machholz I, as it is also known), which appears to be the parent body of the entire complex. Not surprisingly, the system has become known as the "Machholz Complex."

According to comet expert Maik Meyer, the position reported by Airy is not *quite* in agreement with a Marsden comet, but it is quite close and is consistent with an outlier of the group or an object within the broader complex. Unfortunately though, the meager records do not allow the motion of the object to be determined and without this, any further speculation as to a relation with the complex is no more than an intriguing possibility.

By the way, comets of the Marsden Group can make very close approaches to Earth, and some may even by capable of colliding with us. But don't worry. These are such small and fragile objects that they would almost certainly break up into very small fragments after entering the atmosphere.

Despite the trickle of observations of tiny round spots on the Sun and unidentified "stars" seen at solar eclipses, no proof of "an intra-Mercurial planet" managed to stand the test of time. Transiting black dots could never be satisfactorily distinguished from small round sunspots, and the bright objects allegedly seen near the Sun during the 1869 and 1878 eclipses failed to show up in more detailed searches made during subsequent events.

The strongest evidence, in the absence of definitive observations, remained the departure of Mercury's orbit from strictly Newtonian mechanics. At the beginning of the twentieth century, the only explanation appeared to be the existence of a significant mass between Mercury and the Sun, but whether this was in the form of a single planet (which appeared less and less likely as the years passed), a ring of asteroids, or even cosmic dust (possibilities which Leverrier himself raised at the very beginning of the Vulcan saga), was debatable.

Then, in 1916, Albert Einstein published his Theory of General Relativity – in effect, a new theory of gravity superseding that of Newton. When Mercury's orbit was computed using the new theory, it was found to fit perfectly with observational data. The problem of the planet's orbit had finally been solved, not by the discovery of a new planet but by the discovery of a new physics!

How did Einstein's new theory solve the problem?

To answer this, let's take a look at what the problem was really about. What, after all, was so wrong with Mercury's orbit to cause all this fuss about planets hugging the Sun?

Briefly stated, the point in the planet's orbit lying closest to the Sun (perihelion) slowly drifts. In a sense, Mercury does not come to quite the same point in space at each orbit, although the difference is so slight that it takes many orbits for this to become apparent.

As it stands, that is not necessarily a problem. Thanks to the gravitational perturbations of the other planets and even the slight deformation in the Sun's shape caused by its rotation, the perihelion of Mercury was predicted (on strict Newtonian theory) to shift slightly. In fact, the perihelia of *all* the planets show this effect, which is technically known as *precession*.

The problem was, Mercury's rate of precession was too large. Using Newton's Theory and taking all known perturbing influences into account, Mercury's rate of precession was predicted to be 5,557 s of arc per century. But the actual measured value turned out to be 5,600 s of arc per century. The discrepancy of 43 s of arc per century, though it might not seem very large, is still far too great to be brushed aside, and the most reasonable explanation prior to Einstein seemed to be the gravitational influence of some massive object or objects orbiting on the inside of Mercury's orbit.

Einstein's Theory of General Relativity, however, predicted the planet's precession rate exactly. The reason has to do with the way in which a massive body such as the Sun modifies the shape of the space-time continuum. To use a two-dimensional analogy, it is like taking a flat piece of paper, cutting out a narrow wedge from the center, and then rejoining the sheet to make it slightly cone-shaped. If an ellipse had been drawn on that sheet of paper, we would find that it no longer joins up correctly. If continued, it would not repeat until slightly after going all the way around the periphery. The "meet-up point," in effect, would have moved. Although impossible to picture, a 4D version of this (in three spatial dimensions, plus one temporal) is what Einstein's equations describe.

Being deep within the Sun's gravity well and having an unusually eccentric (strongly elliptical) orbit, this relativistic effect is quite pronounced for Mercury, whereas in the case of the other planets, it is too slight to have made their departure from Newtonian predictions obvious.

Finding a solution to the Mercury problem in terms of revised theory did more than eliminate the need for a large mass between that planet and the Sun. It effectively placed an upper limit on the amount of matter permitted to reside there. In other words, if the motion of Mercury as it orbits the Sun can be completely explained without recourse to the pull of unseen matter, the quantity of matter inside its orbit must be too small to have any significant influence on its motion.

That does not necessarily mean that *no* matter exists there. It just means that if anything does lurk there, its total mass must be much, much, less than that derived for the classical Vulcan.

But Is There Room for Mini Vulcans?

During the last half of the twentieth century, the Vulcan idea was revived in a greatly modified form, most notably by Henry C. Courten of Dowling College, Oakdale, New York. Courten had photographed several total solar eclipses since 1966, and some suspicious images had been noted. The most striking results came from the eclipse seen in Mexico on March 7, 1970, when at least ten faint unidentified images were recorded on Courten's photographs. Some of these may well have been artifacts on the photographic plate, but Courten felt that at least seven were real objects. He based this conclusion on the results of sensitive computer analysis, crosschecking images on separate plates. Moreover, some of the objects were apparently confirmed by another observer in North Carolina, and one image seemed to correspond to an observation made by a third person in Virginia.

With the launch of *Skylab* in 1973, Courten hoped to find confirmation of his suspicion that an asteroid, or even a belt of small asteroids, orbited the Sun at a distance of about one quarter the diameter of Mercury's orbit. As a grant investigator of *Skylab* white-light coronagraph data, he searched for telltale images, but without success.

Since *Skylab*, more sensitive coronagraphs have been launched into space, but no "vulcanoids" (as hypothetical small intra-Mercurial bodies have been named) have as yet been recorded. The SOHO spacecraft has been examining the Sun and near-Sun space since 1996, regularly recording stars near the corona as faint as those seen close to the limit of 7×50 binoculars on a dark night, but nothing orbiting between Mercury and the Sun has turned up. More recently, since early 2007, the twin STEREO spacecraft maintain an even finer scrutiny of the Sun and its immediate environs. Stars down to around 250 times fainter than the dimmest seen by naked eyes on dark nights are accessible to this project, but as yet no vulcanoids have turned up in the data.

What these extraterrestrial coronagraphs (most notably SOHO) *have* found, however, are myriads of tiny comets. As of mid-2008, SOHO had chalked up some 1,500 of them, sometimes several being visible in the same image. Many of these appear as simple points of light or small disks, and it is very possible that similar objects could have accounted, not just for Courten's images, but also for earlier "Vulcan" sightings. As already remarked, the object found by Airy and Swift seems particularly open to this explanation, as Airy himself admitted.

This is not to say that nothing apart from comets venture inside Mercury's orbit. The asteroid Icarus has long been known to fly closer to the Sun than this planet, and several other Sun-approaching asteroids have been found in more recent years. But none of these remains within the orbit of Mercury and as such cannot qualify as a genuine Vulcanoid.

So are there any Vulcanoids?

Such things probably do exist, but it is also likely that they are very small and comparatively few in number.

Mercury is a small planet with a weak gravitational field, and its cratered surface bespeaks of many impacts in ages past. Even today, while only a small number of asteroids continue to cross its path, large impacts must occasionally happen. Sizable chunks of rock presumably get hurled out into the surrounding space every now and then. Just as chunks of our Moon are known to orbit close to the Earth/Moon orbit, similar rocks probably trail Mercury around the Sun. Some of these may well orbit on the solar side of the planet. In addition, some asteroid fragments may have migrated into intra-Mercurial orbits from further out in the planetary system.

By definition, such would be Vulcanoids, even though they may only be a few yards in diameter and very, very, faint.

Maybe someday, they will be found.

Bright Lights in the Sunshine

Not every mystery object seen close to the Sun was assumed to be Vulcan. Over the years, some truly brilliant points of light have turned up very close to the Sun and shone so brightly as to be seen with the unaided eye in very bright twilight or even in broad daylight. Far too bright to be confused with an intra-Mercurial orb, these observations belong in a different category from the relatively faint and elusive bodies of the previous section.

Reports of bright star-like objects seen close to the Sun in day-time hours go back to before the time of Christ. Typically, these are no more than brief mentions of "a star was visible during the day" or something similar, and it is not always clear whether the object was a fleeting meteor, a sighting of the planet Venus by day, or something else.

For instance, what can be said about the "blazing starre seen near unto the Sonne" on Palm Sunday in the year 1077? Eighteenth century astronomer G. Pingre opined that this may have been Venus near inferior conjunction, while others suggested a comet very close to the Sun. A meteor is not likely in this instance, and the possibility of a bright nova is remote. Galactic supernovae are probably the least likely explanations for any of the day stars, as the extended duration of these titanic stellar explosions is too great to explain something seen so briefly.

In more recent times, we have the strange instance of a number of folk at Broughty Ferry in Scotland seeing a star-like object not far from the Sun on December 21, 1882. According to the witnesses, this "star' had "a milky appearance" with the naked eye, in the sense that it was not bright and clear like a star seen during the night. Through a "glass," the object was said to have been crescent shaped. Opinions soon divided as to whether this was our old friend Venus or a bright comet near the Sun. By the way, 1882 saw two comets visible in full daylight close to the Sun (one telescopically, the other naked eye) and a third visible almost within the corona during a total solar eclipse. The suggestion that the Broughty Ferry object was of a similar nature seems quite logical.

On the evening of June 26, 1915, Anna Caroline Brooks (daughter of famous astronomer William R. Brooks) noticed a very bright point of light about 5° above the spot on the western horizon where the Sun had set just 10 min earlier. She compared its brilliance with that of Venus (although it certainly was not that planet) and pointed it out to three of her companions, who also had no trouble seeing it. After about 2 min, a cloud covered the object, and it was not seen again. Both Anna and her father searched for it on the following evening, but to no avail.

Soon after his (negative) observation on the second night, William Brooks wrote a brief report in *Popular Astronomy* placing the observation on record "in view of future developments," but, alas, there appear to have been no such "future developments," and Anna and her three companions remain the object's only witnesses. Brooks himself suggested that the object was probably "the nucleus of a bright comet, the tail being invisible from the overpowering light of the sky."

Just 6 years after the Brooks' sighting, there occurred one of the best known of all "bright objects near the Sun" events. This one was especially interesting in so far as the principal observers included gentlemen who must be numbered among the more prominent astronomers of the day.

Late in the afternoon of August 7, 1921, a group of people including Professor H. Norris Russell, Major Chambers, Captain Rickenbacher, Lick Observatory Director Professor W. W. Campbell, and Mrs. Campbell were sitting on the porch of the Campbells' residence at Mt. Hamilton watching the setting Sun when, just as the Sun was about to sink out of sight, Major Chambers spied a bright star-like object a little to its left. Captain Rickenbacher then admitted that he had been watching the object for several minutes, but presumed that it was something known and had not bothered mentioning it. Professor Campbell then retrieved a pair of binoculars from inside the house and managed a quick glance of not more than 2 s before the "star" set behind a layer of horizon cloud. As well as Campbell could determine, the object appeared equally star-like through the binoculars. The object clearly shared the diurnal movement of the sky and must therefore have been a true astronomical body rather than something in Earth's atmosphere catching the waning sunlight.

Although the Lick observers made careful searches on following nights, the mystery object was not relocated. To place the incident on record and, hopefully, elucidate sightings from elsewhere, the observation was distributed telegraphically on August 8 and printed in a *Harvard Observatory Bulletin* (at that time, the official announcement circular for astronomical discoveries) on the ninth.

Not unsurprisingly, these announcements did indeed uncover several sightings of bright objects deep in twilight or in daylight during early August. Equally unsurprisingly, nearly all of them could be explained as Jupiter or Venus. A report from Germany, for instance, if actually referring to the same object, suggested a motion of 27° in just 8 h, placing it as close as twice the distance of the Moon! On closer examination, however, the position of the German object came in very close to that of Jupiter, and there can be little doubt that it was, indeed, the planet that was seen.

Two observations from England looked more promising. Also around sunset on August 7, but 7 h earlier because of the difference in longitude, a bright object was noticed by Lieutenant F. C. Nelson Day and several other people at Ferndown in Dorset. Around the same time, a Mr. S. Fellows also noticed it from Wolverhampton and noted that through binoculars it appeared reddish in color and elongated in the direction of the Sun. Fellows estimated its distance from the Sun as 6° while the Lieutenant gave it as 4°. Despite this discrepancy, both observers made it somewhat further away than the Lick group, who measured it as distant "three degrees east, one degree south" of the Sun. Also, whereas this group estimated its brightness as "brighter than Venus," Nelson Day's estimate was closer to the brightness of Jupiter, although too much weight shouldn't be placed on this.

In the initial announcement from Lick Observatory, the object was suggested to have been the "nucleus of a bright comet, less probably a nova." According to J. A. Pearce, writing in the *Journal of the Royal Astronomical Society of Canada*, the distance from the galactic plane "would almost certainly rule out" a nova. We might also add that the coincidence of having one of the brightest nova on record (apart from galactic supernovae, which it certainly was not) appearing almost directly behind the Sun is highly improbable, to say the least. Moreover, no star in the vicinity shows a light echo or gives any evidence of having recently been an unusually close nova.

On the other hand, a comet is very probable, as it is for the similar but less well observed object of 1915. Comets brighten as they approach the Sun, and one reaching a very high brilliance in its immediate vicinity is not at all unusual. Moreover, the hint of movement between the English and Lick observations are consistent with a comet moving toward the Sun and the object's apparent elongation as reported by Fellows might hint at dust being swept back from the nucleus and into the tail (the latter being invisible against the bright sky). On the other hand, the departure from a star-like appearance may have indicated nothing more than imperfect optics in Fellows' binoculars!

Why were the purported bright comets of 1915 and 1921 not seen in the night skies?

Picture a relatively small comet approaching the Sun from the region of the Solar System opposite Earth. From our perspective, it would stay close to the Sun in the sky (i.e., deep in twilight) and beyond it. If the comet was not especially bright intrinsically, it would be difficult to find against the bright sky.

Suppose, further, that the comet moved in an orbit that brought it very close to the Sun (well within the orbit of Mercury?) and that, at its closest approach, it actually whipped around more or less in front of the Sun – though not *directly* in front – as seen from Earth.

Now, we know that comets normally brighten greatly as they approach close to the Sun, but another very interesting effect may have come into play as well. Have you ever noticed pieces of thistle-down and lengths of spider web (and spiderlings themselves for that matter) brightly illuminated as they pass in front of the Sun on a clear and windy day? Within a few degrees of the Sun, they shine bright and silvery, but fade to invisibility as they move just a short distance away.

This is an example of a phenomenon known as forward scattering of sunlight, and it applies as much to the dust particles surrounding a comet's nucleus as to thistle-down and spider webs blowing about in our atmosphere. In fact, if a dusty comet passes very close to the Earth/Sun line of sight, it is possible for its brightness, as seen from Earth, to increase by a factor of several thousand. But only while the Earth/comet/Sun angle is large. Although the effect peaks near 180°, it already becomes detectable at angles of about 110°.

Now, back to the objects of 1915 and 1921.

Assume that these objects were, indeed, comets, and further assume that each spent only a short time on the earthward side of the Sun. This geometry would have occurred about the time the comet was passing closest to the Sun itself, and the combined effect of this proximity and forward scattering may well have caused its apparent brightness to surge by at least several hundred fold, albeit for only a short period of time (the actual length of time depending to a major degree on how close the comet actually approached the Sun). From obscurity, it suddenly blazed out in full brilliance, only to whip back behind the Sun again and out of forward-scattering geometry. With brightness falling as fast as it surged and the comet pulling away behind the Sun, it rapidly faded to obscurity in the twilight.

This may be the most likely explanation for the 1915 and 1921 objects, and maybe for that of 1882 as well. But it was clearly not the explanation for a bright object seen during the total solar eclipse of June 30, 1973, which we will take a look at now.

From an observing site in Kenya, several photographs taken by more than one camera recorded what appeared to be a bright star-like object close to the eclipsed Sun. It was estimated to have been almost as bright as Jupiter and apparently remained stationary through the time taken to secure the photographs.

From that information alone, it may appear justifiable to conclude that this was yet another instance of the type of object seen in 1915, 1921, and possibly 1882, presumably either the nuclear region of a comet near the Sun or a very bright nova. Because it appeared close to the Sun and because no fading nova was recognized when that region of sky later became visible before dawn, a comet would once again seem the most likely culprit. This seems a logical conclusion – except for one small problem. Photographs taken from other sites during the same eclipse failed to show any sign of the mystery object!

Its appearance on multiple photographs taken with more than one camera presumably established it as a real object and not simply a ghost image or such like. But if it was only visible from a single site, it can hardly have been at astronomical distances.

What, then, could it have been?

Really, we don't know. Someone suggested that it may have been a weather balloon high in the atmosphere. This suggestion, or some close variant of it, is probably about the best that we can do. Some reflective object floating in the air and catching the light of the Sun's corona may indeed explain this mystery object.

On that somewhat unsatisfactory note, we leave the realm of mystery objects near the Sun and direct our gaze in the direction of our neighboring planets – into the realms of Mars and Venus and the odd and interesting things that have been reported there over the years.

3. Planetary Weirdness

Over the years, odd things have from time to time been reported on the surfaces of our neighboring worlds. Traditionally, these reports have come from observers peering visually through the eyepieces of telescopes, although in more recent times some curios have been gleaned from spacecraft transmissions close to the planets themselves.

Mysterious Mars

Mars has been the main focus of these curious events. Of all the planets, it is the one that reveals the clearest "landscape" to the telescope, and it has also been the one associated most often with the possibility of extraterrestrial life. It is therefore not too surprising that it has chalked up quite a list of anomalies, and that many of these have been used at some time or other as evidence for Martian life.

It is worth pointing out that the readiness to jump to the "life" conclusion is in itself a bit of a curiosity. Scientists who remain skeptical of this issue are often portrayed as killjoys who go out of their way to disprove the existence of life elsewhere. Nothing could be further from the truth. Most scientists wish that incontrovertible proof of extraterrestrial life would turn up, but wishing does not make it so!

There is a rule of scientific method known as "Ockham's Razor" (sometimes spelled "Occum's Razor"), which effectively says that the simplest explanation for a phenomenon is the one most likely to be correct. This rule is named for the medieval philosopher William of Ockham (c1270–1349), although it has not been found explicitly stated in any of his known writings. In any case, it is not an invention of scientists but a property of nature. Or, at least, a recognition by scientific method of a property of nature.

Nature typically seems to take the simplest route. British philosopher Bertrand Russell called this the "Law of Cosmic Laziness," but the present writer thinks the "Law of Cosmic Efficiency" is more respectful and sounds nicer. But whatever we call it, it means that in the natural world, things happen with the fewest complications. Ockham's Razor is simply the acknowledgement of this.

So what has this brief excursion into philosophy got to do with the present subject?

Simply this. Life is the most complex known phenomenon in the universe. It is therefore (on the above-mentioned principle) the least likely explanation for any observed phenomenon. In fact, it should be offered as an explanation only as a last resort, when everything else fails.

The observational history of Mars has been a great example of the string of false alarms that happen when this is forgotten. Streaks seen on the surface of the planet could be caused by chance alignments of isolated features connected together by the pattern-finding tendency of the human brain, long fault lines, or some other feature on the planet's surface ... or canals constructed by intelligent Martians. Which do you think the least likely alternative? Which was the alternative most widely accepted?

Likewise, there are dark areas on the planet that change in intensity during the course of a year. This might be caused by wind-blown dust alternatively covering darker underlying rock during one season and being blown off again as seasonal prevailing winds changed direction. Or, the dark areas might have been vast fields of vegetation. Which one was most popular before the first spacecraft arrived at Mars in 1965? Which one turned out to be correct?

Of course, many people argued that areas on Earth change color with the seasons because of changes in vegetation. But we already know that Earth *supports* life! The complex explanation is acceptable here, but whether Mars has life or not is the whole point at issue. It is an enormous leap of faith to assume that the simple observation of changing hues on parts of its surface warrants postulating the most complex of processes unless there is some very, very good supporting evidence.

Those Notorious Martian Canals!

Although the long tale of the "Martian canals" has been told so often that there may seem little need to reiterate it here, a brief account of the saga sets the stage for many of the early (and not so early!) speculations about Mars.

The very idea of "canals" would not even have raised its head had there not been a fairly widespread belief that liquid water existed on Mars and remained sufficiently stable to flow over long distances across the Martian surface. This belief appeared best supported by the so-called "Lowell band." Not the name of a musical ensemble, the "band" refers to a thin bluish rim that Percival Lowell and just about anyone who seriously observed Mars through a telescope found bordering the shrinking ice caps during the Martian spring. Lowell, quite logically, explained the band as evidence for liquid water – or marshy ground at the very least – lingering in the wake of the ice cap's seasonal retreat. Assuming the ice to be frozen water rather than some other material, it seemed reasonable to assume that melt water might well be left as it thawed.

Then there were the dark areas, such as the very conspicuous triangular feature known as Syrtis Major. The first telescopic observers of the Red Planet thought that these regions were oceans, but this explanation was abandoned as long ago as 1863, when G. V. Schiaparelli pointed out that they do not reflect the Sun's light in the way that bodies of water should. But with the demise of the oceanic model, a new and even more exciting possibility emerged: The dark regions were vast fields of vegetation!

Many people saw this hypothesis as having much in its favor, with or without Ockham's Razor. For one thing, the features were not entirely static, unlike the dark areas on the Moon. Dark formations such as the romantically, if inaccurately, named Solis Lacus or Lake of the Sun, altered in shape and at times even appeared to break up into several different sub-regions. Dark areas were even known to emerge from the ochre "deserts," simply appearing where no dark region had been previously noted.

Then there were the color and hue changes noted by many astronomers and, following the advent of planetary photography, apparently confirmed by the photographic plate. Many astronomers claimed that these changes were seasonal (apparently supporting the claim that some form of plant life was responsible), although there has never been complete agreement on this score. Thus, whereas E. M. Antoniadi maintained that Syrtis Major changed from bluish-green in winter to brownish in early summer, Charles Capan described that region's color change simply as a change from blue-green to green-blue. Other astronomers, for example Sir Patrick Moore, admitted that they had never detected any definable coloration at all in the dark areas, let alone anything that could pass for seasonal variations!

The same could be said for the "wave of darkening" allegedly witnessed by some observers. As the polar cap shrinks, a progressive intensification of the dark regions was said to spread toward the equator as a sort of advancing front. Other equally skilled observers of Mars (Sir Patrick Moore is a good example) did not see any such thing. Changes – yes. Maybe even slight seasonal variations in intensity and/or hue. But nothing as regular and orchestrated as a "wave of darkening."

PROJECT 9
Color Changes on Mars*

Here is something for experienced planetary observers.

What is your impression of the dark areas on Mars? Observe them when the planet rides high in the sky and Earth's atmosphere is steady, use a reflecting telescope (of at least 8 in. or 20 cm aperture) to minimize any false color estimate.

Do you agree with those observers who reported tinges of green or some other discernible coloration of these regions? Or do they simply appear dark but non-descript?

You might like to keep a record of your impressions of the these regions over time. Do they appear to darken or change color as the Martian seasons change? Can you find anything that you might describe as a "wave of darkening" as the Martian summer progresses? From your observations, do you suspect that some of the earlier reports may have been influenced by the observers' subconscious belief that fields of vegetation were being observed? Or do you conclude that real colors are discernible and genuine seasonal changes do occur?

Of course, the big problem with all of this is the very small size of the Martian image in a telescopic eyepiece. All of these features are, to a greater or lesser degree, best described as "tiny," and the hues marginal. Moreover, the sensitivity of people's eyesight differs. One person might detect something that another fails to see. Conversely, simple differences in contrast might cause one person to "see" something that is not even there.

Yet, when all of these observations of the Lowell band, the wave of darkening, and variation of dark regions were put together, the case for vegetation looked pretty convincing to most astronomers and just about all astronomy popularizers. Many astronomers and probably the majority of laypeople prior to 1965 found this combination a strong argument in favor of Martian life.

The scenario went like this:

1. As the polar ice caps melt, first of all liquid water spreads to the surrounding regions, not necessarily as an actual fluid sheet but at least as wet soil and possibly marsh-like conditions.
2. Because the Martian air is so thin, liquid evaporates quite quickly, and a cloud of humid air spreads towards the Martian tropics, reviving thirsty Martian plants in the dark areas as it goes. Earlier speculation even suggested that liquid water might have flowed away from the melting ice caps (as we will see shortly), but as the tenuous nature of the Martian atmosphere became better appreciated, this looked decreasingly likely.

Not everyone was won over, of course. The Lowell band came in for special criticism. For instance, Antoniadi wrote in 1930 that the band was actually an optical illusion arising from the contrast between the white polar cap and the less reflective areas immediately adjacent to it. He pointed out that the band "does not obey the laws of perspective" and noted that it does not show in photographs, proof (in his opinion) against its objective reality.

Others, as we have remarked, denied the reality of the wave of darkening.

Nevertheless, before skepticism concerning one or other of these matters began appearing, the idea of a vegetated Mars watered by the seasonal melting of the ice caps looked very reasonable, and it was against this broad background that the controversial saga of the famous – or infamous, depending upon your point of view – canals developed.

The saga officially began in 1877 when Schiaparelli recorded, during the favorable Martian opposition of that year, some 40 thin, straight features crossing the ochre-colored "light" regions of the planet. This was not actually the first time that "canals" had been recorded. A single canal-like feature appears on one of the maps drawn by W. Beer and J. H. Madler in the 1830s, and a drawing by Rev. W. R. Dawes in 1864 shows features that would certainly have been called "canals" post 1877. Yet, it was Schiaparelli's observations of that year that placed these features – for better and for worse – squarely in the thick of Martian research and the public mind alike.

The actual word that Schiaparelli used was "canali," which is not precisely equivalent to the English "canal," with the latter's implication of intelligent construction. It is closer to "channel," which may or may not imply artificiality. Schiaparelli appears initially to have suspected the "channels" were simply natural watercourses, maybe flanked by thin strips of vegetation, but their seeming regularity impressed him, and he later came to express the opinion that there was "nothing impossible" in the suggestion being made by some people that they were "the work of intelligent beings."

But it was Percival Lowell who chiefly took up the idea of the canals as a vast system of engineering constructed by a race of highly advanced Martians.

Looking back from our era, we may see Lowell's speculation as an example of just how farfetched things can become when Ockham's Razor is kept in its sheath. Yet, in Lowell's day, the ideas he put forward did not seem as way out as they do to us. After all, little was known about the Red Planet at the time, and, as we have seen, the little that was known seemed quite amenable to the belief in plant life, at the very least, on our neighboring world. From this, the step to animal life and thence to intelligent animal life did not seem too great.

Lowell's romantic planetary perspective saw Mars as a drying and dying planet and its intelligent inhabitants as being knitted

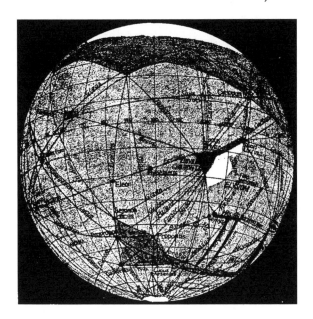

Mars as Lowell pictured it: A planet networked by canals!.

together into a global society struggling against the inevitable des-
iccation of their world. The Martians of Lowell's imagination had
succeeded in doing something that we fighting and factionalized
humans have still not accomplished. But then, their plight was
more desperate than ours!

Lowell's ideas about Martian life were more extreme than
those of most of his contemporaries, and his hypothesis concerning the nature of the canals became increasingly isolated as more
was learned about the planet. Yet, the belief in life of some sort
on the Red Planet was almost universal prior to 1965 (the year of
Mariner 4's great shock), and even as late as 1962 a book was published arguing for intelligent life on the planet (and, incidentally,
on Venus as well – a far harder thesis to uphold as late as 1962).

It is against this general background that any odd happening
on Mars tended to be interpreted.

But what of the canals?

Besides a few linear features, most turned out to be more in the
eye of the beholder than on the surface of the planet. The human
eye/brain combination is pretty good at connecting dots, and that is
essentially what was happening. Isolated and unconnected features

on the Martian disk observed close to the limit of perception were strung together by our innate ability to create patterns in what is essentially random noise. The Martian canals were indeed intelligent artifacts, but the intelligence was behind the eye of their observers!

Mars Calling?

On the night of July 28, 1894, M. Javelle of the Nice Observatory noticed a luminous point just outside the illuminated portion of the gibbous Martian disc. It seems that the object was, nevertheless, within the boundary of the total disk and not simply something in the background or foreground (an asteroid perhaps?) observed almost in line with the planet's limb. This latter possibility, however, could not be totally excluded.

Writing in response to this observation, an anonymous contributor to *Nature* expressed the opinion that, because the luminous point must have "[either] a physical or human origin" it could be expected that "the old idea that the Martians are signaling to us will be revived."

This anonymous author then suggested some possible (natural) causes for the event, namely, aurora (which he thought unlikely unless Martian aurora are much more intense than terrestrial ones), a long range of snow-capped hills, or "forest fires burning over a large area." Note that one of these suggested "natural" explanations required at least vegetative life on Mars.

Whatever his private thoughts were, he then noted that "Without favoring the signaling idea" a more favorable time for sending such a signal "could scarcely be chosen." His reason for saying this apparently rested on the gibbous phase of the planet, causing some of the disk to remain in darkness. Apparently, Mars was close enough to Earth at the time for such a "signal" to be visible, yet during its *closest* approach at opposition, the disk would appear fully illuminated and any signal likely lost against the bright background.

The *Nature* article concluded that "The Martians, of course, find it much easier to see the dark side of the Earth than we do the dark side of Mars, and whatever may be the explanation [of the luminous phenomenon] … it is worth pointing out that forest fires over large areas may be the first distinctive thing observed on either planet from the other besides the fixed surface markings." From this, it may appear that the author leaned toward the forest fire explanation, but could these have been deliberately lit as a signal? The answer probably depends upon the level of environmental awareness among Martians! Realistically, the luminous spot was probably nothing more than a high-altitude Martian cloud catching sunlight, something the author of the *Nature* article apparently failed to consider but which (let it be noted) was the favored explanation of Percival Lowell in a paper about this and similar observations published in 1901.

Early that year, word was sent out that an astronomer at Lowell Observatory had seen a "shaft of light" beaming up from a location on the surface of Mars, sometime in early December 1900. In a follow-up announcement, this observer was described as being careful and reliable, although his name was not given. The shaft of light was said to have persisted for "seventy minutes." The initial telegraphic announcement of this event caused a flurry of wild ideas, including a story circulating through Europe that Professor W. H. Pickering, of the Observatory, had actually been in communication with Martians!

In the 1901 paper mentioned above, Lowell pointed out that the original description of this event was that of a "projection" over Icarium Mare. The sense of "projection" apparently became distorted as the report traveled. Lowell thought that the phenomenon was due to a cloud rising high into the Martian atmosphere over the Icarium Mare, which, by the way, he interpreted as "a great tract of vegetation".

It is interesting to note that Lowell, in spite of his strong belief that Mars was inhabited by intelligent, technological, beings, was also commendably objective in his assessment of the temporary luminous spots. He did not jump to the quick conclusion that every unusual happening on Mars is somehow associated with life. He knew how to shave with Ockham's Razor! Temporary phenomena similar to the above have been reported on other occasions as well, but it would be tedious to recount each incident as, surely, the explanation given by Lowell (or some close variant of it) is the correct one.

One further example of a bright spot will be mentioned, however, more as an example of people's reaction than for additional information on the phenomenon itself. This was the one observed by Japanese astronomer Tsuneo Saheki in 1951 and which caused considerable comment all over the world at that time.

Remember that the year 1951 was a time of change, uncertainty, and fear. The world had quite recently been through the worst war in history; a war that ended with the detonation of an entirely new and unpredictably dangerous form of weaponry. The communist and non-communist worlds faced each other with renewed suspicion and hostility now that the common foe of the fascist demon had been exorcised. On top of this, pilot Kenneth Arnold had recently reported seeing a number of mystery objects moving through the sky with a motion that he described as being "like saucers skipping over a pond," or words to that effect. Although he said nothing about these objects being saucer-shaped, the term "flying saucer" was quickly coined, and the wild assumption that they were extraterrestrial spaceships gained currency among a growing segment of the public.

This was the background against which a bright and short-lived spot of light suddenly appeared on Mars!

Over a decade later, Patrick Moore (later Dr. Patrick Moore and, still later, Sir Patrick Moore) recalled in his book *The Amateur Astronomer* how a national newspaper telephoned him one morning at 2 a.m. to ask for his views about "the atomic bomb that had gone off on Mars." Moore did not record his reply!

The old notion that the Martians might be sending us a signal revived again in some minds, while some flying saucer true believers were probably just as convinced that yet another Martian spaceship had taken off for Destination Earth.

Moore himself basically took the Lowellian line that the spot was a high Martian cloud catching sunlight, although he also offered the suggestion (which he nevertheless considered doubtful) that it might have been a volcanic eruption. The impact of a meteorite was also suggested by some; however, the correct explanation, in line with the earlier incidents of bright Martian spots, is most likely that offered long ago by Lowell.

Bright spots are not the only temporary Martian markings that some people have suspected as being artificial. In the January 1926 issue of *Scientific American*, controversial astronomer William Pickering wrote that "It is a rather curious coincidence that at each of the recent very near approaches of the Earth to Mars, strikingly regular, although only temporary, geometrical figures should have appeared upon its surface." He goes on to tell of the "well-known cross" that appeared during the 1879 opposition and which was recorded by no less an observer than Schiaparelli. This cross apparently materialized at the center of the light-colored circular feature known as Hellas. At the previous opposition 2 years earlier, that same observer saw no such thing within the circle of Hellas, only a single and previously known vertical "canal." During the oppositions following that of 1879, the cross was replaced "by an irregular curved structure," according to Pickering.

Then in 1892, a pentagonal figure some "800 miles" in diameter appeared in the Arequipa region. The pentagon was apparently centered on a small dark feature that was later given the name of Ascracus Lake.

The year 1924 saw the closest approach of Earth and Mars in the twentieth century, and Pickering noted that the opposition was marked by "an unusually large and complicated figure." As in 1892, this new figure was also a pentagon, except that this one was twice as large as the former. Pickering remarks, a little tongue-in-cheek we may presume, that the pentagon is "apparently a favorite figure with the supposed Martians" and briefly recalls yet another pentagon "well known to all students of the planet" that once appeared in Elysium. This figure – about the same size as the one in Arequipa – subsequently transformed itself into a circle!

PROJECT 10
Seeing Fine Details on Mars*

A question for experienced observers of Mars.

What do you think of the claim that geometric figures of the size spoken about here could be recorded on Mars? Using the telescope with which you are most familiar, try to estimate the size of a feature on the planet that you could describe with the sort of detail presented here. Based on your own experience, do you think that some of the geometrical regularity reported may have been "in the eye of the beholder" rather than on the Martian surface?

Pickering admits that "it is indeed curious that these complicated figures should occur on Mars." He also found it strange that they are only of temporary duration and that Martian patterns seem to avoid "large four-sided figures" in favor of the more complex pentagon, as well as the circle.

Pickering offers no explanation for these strange figures. He simply states that:

> Some people will doubtless believe that these designs are not due to mere accident, but are artificial, and constructed for our especial edification, and as an announcement of the existence of intelligent life on their planet. If so, we wish the Martians would plant them out, or otherwise construct them, more frequently than once every fifteen and a half years. If not due in the past to chance, we wonder very much what figure will appear at the next close opposition in 1939. However we must not expect too much of the Martians, and if they have been doing this sort of thing for the last 10,000 years or more, we must consider them to be far more persevering in their endeavors to communicate, than the inhabitants of our own self-satisfied, and very unresponsive planet.

Irony and skepticism are mixed here, yet the door on signaling might not be completely closed in Pickering's account.

Nevertheless, his sober assessment of these geometrical markings was in stark contrast to some other ideas vented at times. One observer apparently became convinced that he could read "The Almighty" spelled out in Hebrew lettering on the Martian surface!

In defense of the signaling idea, it is true that geometrical figures do have the appearance of intelligent construction about them, and it might be supposed that the more exact the geometry, the more likely they are to be artificial constructs.

Elsewhere in an entirely different context, the present author coined the term "transitive complexity" to denote a form of regularity whose nature and existence clearly points to some purpose beyond itself. (Anyone interested in following this further is directed to the author's book *Planet Earth and the Design Hypothesis*, where the concept is treated at some length.)

This is not the place to go into this concept in detail, but, very quickly, the argument runs like this: If something appears to have been designed to convey information or to fulfill some other deliberate purpose, and if it does in fact convey that information or fulfill that purpose, it is strongly implied that the thing in question has been purposively constructed for that very reason. One example is a sign pointing to a location down a road actually leading to the signified destination. A blueprint of a functioning electric motor is another example.

In the Martian context, Pickering's figures do not obviously show transitive complexity. For them to do so, they would need to convey verifiable information.

We can think of some possibilities that hypothetical Martians might use to arouse the interest of Earthlings.

Mathematics is the same on Mars as on Earth, irrespective of what symbols might be used to express it. So a thoughtful Martian could construct the following pattern:

???

The simple arithmetic pattern here is pretty obvious, and if something like *that* appeared on Mars, astronomers would be interested. Very interested indeed!

Another possibility would be the shape of a right-angled triangle with squares constructed on its three sides. If the squares were truly accurate, the area of the one on the hypotenuse would equal the sum of those on the other two sides. On Earth, we call this Pythagoras's Theorem, but it holds equally on Mars, where it might be known as Xenrs' Theorem (or whatever Martian word "Theorem" translates into).

Of course, nothing in the actual markings suggested anything of the sort.

As a final word on the markings, we must not forget that Mars never looms large in our telescopes. Even during the exceptional opposition of 1924, its disk was still little more than 25 min of arc in diameter. On that scale, the very largest of the pentagons was little more than 9 s of arc across. As most of the markings were smaller than this, and most oppositions of the planet less favorable than that of 1924, it is clear that these features are anything but large and conspicuous.

The human brain is good at picking out patterns. That is part of its job, but that also means that patterns seen marginally can be exaggerated and spurious ones can be constructed out of white noise alone. Observing so close to the limit of discernment, markings probably were "seen" as more precise and geometrical than they really were. Coupled with a desire (even if only subconscious) to find evidence of intelligent life on the Red Planet, this might just have been enough to transform very crude geometrical patterns – of a type entirely compatible with the interaction of wind and other natural forces – into something looking suspiciously like intelligent signals.

Patches, "Varnish," and Gillevinia Straata?

Pre-space age observations may have uncovered some odd features on Mars, but it was only when images started coming back from unmanned craft and from landers on the actual surface of the planet that the true nature of Mars became apparent.

But this revelation did not come in a single passage. The first craft, *Mariner 4*, was the real shocker. Prior to the brief series of images beamed back from this craft in 1965, most people thought of Mars as being (more or less) like a scaled down version of Earth. After *Mariner 4*, the perception of the planet changed to that of a scaled up counterpart of the Moon.

As earlier generations compared markings on Mars with features such as seas and vegetation fields on Earth, so similarities between Martian and lunar features were being sought in the late 1960s. This actually met with some success. To the naked eye (and even more through the reverse end of a pair of binoculars), our Moon does look remarkably like a telescopic view of Mars. It even has a few canals, chance arrangements of craters and other surface markings joined together by the human eye. In fact, if Mars really was like an oversized Moon, the markings drawn by Lowell and others began to make sense. Not the sense given by Lowell et al. but sense nevertheless!

PROJECT 11
Mars and Moon

Compare a telescopic image of Mars (or a good quality image or drawing based on telescopic observations) and the Moon as seen naked eye or through the reverse end of a pair of small binoculars. How many features do they have in common? Compare especially the dark areas of both bodies and dark "oases" in otherwise bright regions. Can you detect any "canals" (thin dark lines) on the Moon?

But it was not simply the presence of moonlike craters per se that many found so depressing. The images beamed back from *Mariner 4* looked *so* moonlike that many scientists understandably assumed the surface to be equally ancient. This in turn implied that the planet had never experienced the erosive action of a dense atmosphere and liquid water. If Mars was really as dead as it appeared – geologically and biologically – it probably had always been that way.

The conclusion that the Martian surface was very ancient and unchanging was soon challenged, however. Mars orbits very near the asteroid belt and should therefore experience a higher rate of major impacts than the Moon does. Whether or not that sounds hopeful for life could be debated, but at least it allowed the cratered surface to be not ancient so much as prematurely aged, which in turn implied that erosion had been a lot more effective in the distant past than had initially been thought.

This may have made the pockmarked appearance of Mars a little less shocking, but it did little to lessen the overall impact of the *Mariner 4* data on the public's appreciation of Mars.

Actually, it turned out that the heavily cratered terrain imaged by *Mariner 4* was more moonlike than much of the rest of the planet, and the perception of a basically lunar Mars was not much nearer the truth than the earlier earthlike one. Still, this perception, slightly modified by *Mariners 6* and *7* in 1969, held until *Mariner 9* arrived in orbit around Mars in 1971/1972 and the first close-up global imaging of the planet undertaken. After that, Mars was accepted as being very Marslike!

The popular idea of Martian life, however, took a battering. The harsh cratered landscapes revealed by *Mariner 4* shocked a popular mind that had long been nurtured on visions of vast lichen fields and slender canals. But to the planetary scientist, an even more damaging finding was the thinness of the Martian atmosphere and the fact that it was not composed largely of nitrogen, as had previously been thought, but of carbon dioxide. Scientists had long known that the atmosphere was thin, but even they were surprised to find it as tenuous as it turned out to be – only some 10 mbar at the Martian surface – approximately equal to our own planet's air at an altitude of 15 miles (24 km). And if that was not bad enough, the lack of nitrogen meant that an element essential for life (at any rate, for Earth life) was either rare or missing altogether.

There was a brief burst of excitement in 1969, when *Mariner* 4's successors detected spectral lines near the polar cap that were initially interpreted as being due to methane. This gas is regarded as a possible signature of biological activity, and there was some speculation about possible methanogenic bacteria living at the edge of the polar ice cap. Alas, further analysis showed that the spectral lines were actually due to frozen carbon dioxide – dry ice. The polar cap was not even frozen water but solid CO_2! (As we shall see, however, that was far from being the last we heard of methane!)

Mariner 9 and, even more so, *Viking 1* and *2*, which sent landers to the surface of Mars in 1976, modified the harsh picture somewhat. This "softer" and somewhat friendlier Mars has been further highlighted by later probes, and the myriad of images of expansive Martian deserts that do not look vastly different from some regions of Arizona and Central Australia, minus the Saguaro cacti and clumps of Spinifex grass, of course!

Moreover, further examination of Mars showed that it is not altogether waterless. Beneath the dry-ice caps lies a large reservoir of water ice.

Mars was the second extraterrestrial body to have images beamed back from its surface, and even the very first revealed a Martian landscape that bore little resemblance to the *Apollo* pictures of the stark and airless Moon. More intriguingly, biological experiments on the *Viking* landers found some sort of active chemistry in the soil that not everyone was (or still is) willing to divorce from a lowly form of life, even though the majority opinion does favor an abiotic explanation.

Two *Viking* team scientists who remain skeptical about the abiotic view are Drs. Gilbert Levin and Patricia Straat. In their opinion, some of the lander results, while not *precisely* what one would expect to be produced by life, did not strictly mimic non-biological (purely chemical) reactions, either.

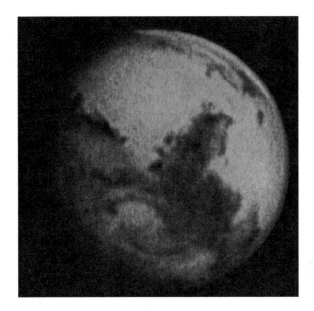

Mars sans canals. NASA image.

A Martian panaroma. NASA image.

The Labeled Release (LR) experiment was their special proj-ect. This experiment involved wetting a sample of Martian soil scooped into chamber within the lander with a nutrient containing the radioactive isotope carbon-14. If microorganisms were present, they would presumably ingest the "labeled" nutrient and expel waste gases marked by the radioactive trace. The presence of life would be betrayed by a slow increase of radioactive gas within the chamber. This experiment gave results that were broadly, albeit not exactly, consistent with what had been expected if life were present. Conversely, no purely non-biological agent has managed, after 30+ years, to yield closely similar results. These scientists and their supporters continue to raise the prospect that, perhaps, the "active agent" in the Martian soil is biological after all!

There is nevertheless one reactive chemical in at least some areas of the Martian surface. The chemical in question is perchlorate, discovered in the Phoenix lander data of 2008. This substance also occurs, though in far smaller concentrations, in very dry environments on Earth, in particular, the Chilean desert. It consists of an atom of chlorine surrounded by four of oxygen. Whether its discovery on Mars strengthens the case for or against life there depends on whose article one reads. Some say that it makes life less likely and may even be responsible for the LR's apparent positive response, while others see it as being something potentially useful for Martian microbes to utilize. Perhaps the safest comment at the moment is that of the Phoenix scientists themselves: "The discovery of perchlorate is neither supportive of, nor detrimental to, the possibility of life in the Martian regolith."

Be that as it may, other recent research could, if confirmed, greatly strengthen the biological explanation for the LR results. In fact, it *has* convinced a minority of scientists, to the extent that one of their number even named the purported organism!

The name? *Gillevinia straata*!

What is *Gillevinia straata*?

Here is a quote from the abstract of a lecture delivered by Gilbert Levin on May 15, 2007, and published in *Electroneurobiologia* Vol. 15.

> **Gillevinia straata**, the scientific name recognizing the first extraterrestrial form ever nomenclated, as well as the existence of a new biological kingdom, **Jakobia**, in a new biosphere – **Marciana** – of what now has become the living system **Solaria**.

The scientist responsible for the naming of this purported organism was Mario Crocco, Director of the Neurobiology Research Center at Ministry of Health in the Argentine Republic. The name itself, suitably, is in honor of Drs. Gilbert Levin and Patricia Straat.

But does Gillevinia straata really exist?

That is the question.

The majority opinion remains negative or, at best, extremely skeptical.

Crocco's naming of the purported organism was met, not surprisingly, by a good deal of protest from the scientific world. Without the "organism" having been seen or its (hypothetical) DNA analyzed, its naming is highly irregular.

Nevertheless, Crocco' supporters point out that this radical move should raise the visibility of the purported organism and encourage further research.

So, do Drs. Levin and Straat have the honor of having their names attached to the first extraterrestrial organism ever found, or have their names been given to a chimera?

Time – and further research – will tell.

Potentially the most convincing finding of this continuing investigation is the alleged presence of a cyclic pattern in the release of gas from the soil used in the LR experiment. In 2001, Joseph Miller found evidence in the *Viking* data of a cycle having a period of one sol (i.e., one Martian "day") and persisting for a period of up to 90 sols. Further papers in 2004 and 2005 by Miller, together with Levin, Straat, and A. Van Dongen, support this result. If confirmed, this cycle would seem to be a Martian counterpart of the terrestrial circadian rhythm. Life – and life alone – shows this phenomenon, and if something similar were found on Mars, it would greatly boost the case for life on the Red Planet. Analysis of the data is continuing in the hope of determining whether this pattern is statistically significant – in other words, a real pattern and not simply something read into random "noise." At the time of writing, it remains a work in progress.

The case for Martian organisms being championed by Levin, Straat, and colleagues would be greatly strengthened if they could find independent evidence that something vaguely describable as "life" grew or crawled about on Mars, and to that end the images taken from the *Viking* landers were carefully scrutinized. In the early 1980s, something that just might have been corroborative evidence was found – small colored patches on some of the rocks and nearby ground, which changed over time. One rock sported a patch that appeared to shift slightly in position, while a faint mark on the soil at the base of another faded and vanished. Were these patches evidence of something growing on Mars?

The question does not appear to have been finally answered to everyone's satisfaction, although other explanations for the patches were soon forthcoming (indeed, Levin and Straat themselves suggested alternatives not involving living organisms). Tricks of the light, even glitches in transmission, may have been responsible for the patches and are considered more likely explanations by most scientists.

More recently, Levin drew attention to a more widespread coating on the rocks shown in *Viking* – and later – images of the Martian surface. This coating appears to be very similar – maybe even identical – to the so-called "desert varnish" or "rock varnish" found in very dry environments on Earth and which is known to harbor rich populations of microorganisms. Indeed, the microscopic life within the "varnish" was for a time suspected of contributing to its very existence. Levin was quick to point out that the discovery of a similar rock coating on Mars may imply biological activity there as well. Recent research has, however, pulled away from the hypothesis that biological activity has an essential role in the formation of rock varnish on Earth. The presence of rock varnish on Mars, even if it does turn out to be exactly the same as its terrestrial counterpart, does not, therefore, necessarily imply that life exists there, although it may well provide a nicely habitable micro-environment for any microscopic Martians that *may* exist.

Nevertheless, Levin's main body of evidence remains the LR results, which he continues to argue support the existence of some type of life on Mars. He criticizes the popular objection that *Viking*'s failure to find organic material in the Martian soil is a fatal one, arguing that the lander's instruments were not sensitive enough to measure concentrations below a certain threshold. Bacterial activity could be perfectly viable at levels undetectable by *Viking*'s instruments, especially if the organisms were very frugal with their waste products, as might indeed be expected in an environment as harsh as that of Mars. In fact, it seems that an amount of organic material equivalent to around one billion bacterial cells could have been present in the sample measured by *Viking* and yet escaped detection. This would allow more than enough organisms to be present to give rise to the reactions noted.

Martian rock varnish? On Earth, rock varnish is a micro-habitat for life. Could that be true for Mars as well? NASA image.

Letters, Faces, Ruined Cities, and Transparent Worms: What Will They Find Next on Mars?

Even if the minority opinion of Levin, Straat, and their colleagues turns out to be correct, and their arguments do end up proving that life truly exists on Mars, the Martian biology in question will bear little resemblance to that pictured by Lowell and most earlier life-on-Mars enthusiasts. The Martians will be little green cells (probably not green, actually!), certainly not little green men. This would greatly excite scientists, but hardly satisfy the "space romantics" and those who still cling to the belief that Mars does (or, more probably, once did) harbor intelligent beings of some sort. If this latter belief is going to stand, some very radical evidence needs to be found, preferably in the form of ancient artifacts on the surface of the planet.

Surprisingly – or perhaps not! – the more adventurous advocates of advanced Martian life have from time to time managed to find some such "evidence" to their liking, such as the markings seen through Pickering's telescope. Certain features turned up in *Viking* and post-*Viking* imagery that on first glance looked surprisingly artificial. Naturally enough, the evidence turned out to be, on more thorough investigation, even more tenuous than the Martian air, but the story remains an interesting one.

First of all (and least of all), there were the rock markings.

Although few, if anyone, seriously took it to be an artifact, early in the Viking transmissions a series of marks resembling the letters "2 g b" were found on a Martian rock. Amusingly, 2GB is also the name of a commercial radio station in Sydney, Australia, and the day this was announced listeners were treated to renditions of *The Martian Hop* and any number of Martian jokes. This was all in fun, of course. The possibility of Martians communicating with Arabic numerals and English letters is a bit too far out even for the fringe!

The real Pandora's Box of speculation came not from the *Viking* landers but from the orbiters.

On July 25, 1976, an orbiter image of the Cydonia region of Mars showed one of the mesas in the area bearing a striking resemblance to a human (or maybe humanoid) face. Probably trying to stave off speculation, *Viking* chief scientist Gerry Soffen quickly dismissed it as a trick of light and shadow. Nevertheless, the Face, as it soon became known, had the annoying tendency of turning up on any number of other *Viking* images under different lighting conditions and having different resolutions. And it *always* looked like a face!

In some respects, the Face was to the late twentieth century what the canals had been to the late nineteenth, though fortunately the speculation did not become as entrenched in the public

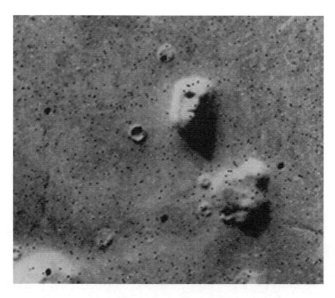

The "Face on Mars" as we initially saw it. NASA image.

mind. But for those who continued to cherish the thought that Mars once harbored a civilization, the Face was welcomed with open arms.

Moreover, as enthusiastic eyes scrutinized the region around the Face, attention was drawn to a number of interesting formations that looked remarkably like giant pyramids. These even seemed to form part of a wider series of structures that some people interpreted as the ruins of an ancient city. The feature was dubbed "the Inca City" in pro-Face literature. There were even claims that the "city" appeared to be orientated toward the Face!

Articles and books popularizing the Face and its surroundings as evidence for intelligent (past?) life on Mars were published, and the topic became quite a hot one among what may charitably be termed "independent thinkers." The present writer recalls listening to a talk program about the paranormal on a Sydney radio station (not 2GB, by the way!) during which a listener telephoned to talk about the Face on Mars. Although his attitude was restrained, he expressed the opinion that the Face seemed to bear a strong resemblance to the image on the Shroud of Turin. The host hastily agreed, though suggesting that this was getting a little too weird, even for a program about the paranormal. Happily, the listener was of the same mind and the subject was dropped like the proverbial hot potato.

Of course, the "artifact" interpretation of the ruined city, pyramids, and Face was severely criticized by the *Viking* team. One NASA official pointed out that the pyramids were really very crude (his language was actually a bit more colorful than this, but the meaning was clear enough!). Clever Martians should have made a better job of pyramid building, presumably. Others argued that there are formations on our own planet that resemble statues or carved faces but are known to be entirely natural. There was certainly no compelling reason to believe that the Face was any different.

In any event, the issue was finally settled when higher resolution images became available from NASA's *Mars Global Surveyor*, which scanned the planet from 1997 until 2006. The mesa was clearly visible, but the Face faded into a confusion of perfectly natural-looking markings. Like the fabled canals, the Face was also in the eye of the beholder.

By the way, the mesas of Mars are fascinating formations in their own right, Face or no Face!

Amazingly, some of the "mesas" in sub-polar regions are, in actual fact, low formations of dry ice or frozen CO_2! Even more fascinating, these have been found to be shrinking at the rate of some 3 meters (about 10 ft) per year, indicative of a warming Martian climate. What this may mean for global warming on both Mars and Earth is an interesting and inevitably controversial subject, but one which unfortunately lies beyond the scope of this book.

Returning to our topic, we can safely say that the *Mars Global Surveyor* has lain to rest the Face on Mars. Unfortunately, it has also opened another can of worms. Giant glass worms, in fact!

At least, that might have been the conclusion drawn from *Surveyor* images of a Martian gully. The image appears to show a transparent tube about 160 ft wide and several miles in length!

While this could hardly be interpreted as an actual living worm (!), enthusiastic eyes could be forgiven for seeing it as a sort of tube or tunnel constructed for a purpose known only to its

The "Inca City". NASA image.

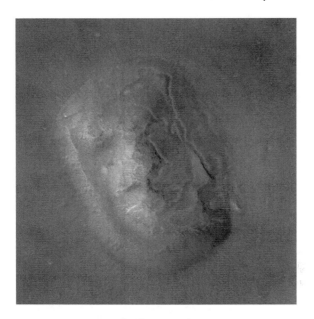

The "Face on Mars" as seen at higher resolution. NASA image.

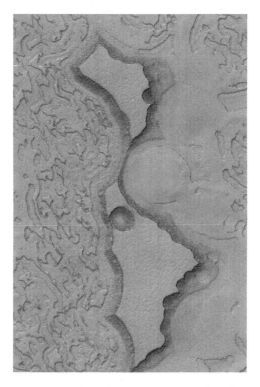

Mesas of Dry Ice; part of the complex Martian landscape fashioned by natural forces. NASA. Courtesy nasaimages.com.

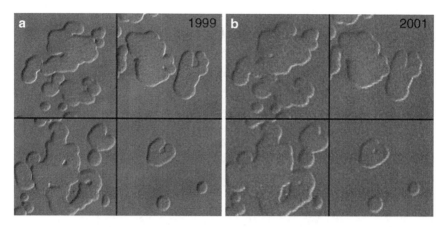

Signs of Martian Global Warming. Depressions enlarge and icy mesas shrink as Martian temperatures warm! NASA/JPL/Malin Space Science Systems Used with permission.

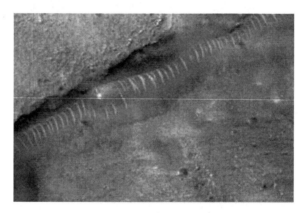

Glass tube, transparent "worm" ... or sunlit Martian canyon? NASA image.

Martian builders. Not surprisingly, that explanation was not long in raring its head.

On the website of Richard Hoagland (long known as a prominent Face supporter), geologist Ron Nicks writes that the "tube" "appears to cling to a desert canyon wall, near the canyon's bottom, and extend along its entire length." He goes on to say that "The feature has the appearance of being 'translucent,' of being supported at somewhat regular intervals by 'ribs,' and of being quite cylindrical – with localized, internal structure at one point of considerably high albedo (brightness)."

A very impressive feat of Martian engineering indeed. Or is it just an optical illusion?

Many of the images transmitted back to Earth from the various Martian orbiters are especially prone to a variety of optical illusion that seems to turn them inside out. You might recall how the first published images of series of canyons actually looked more like series of ridges. In other images, craters become domes.

The illusion can be created and corrected again simply by flipping the pictures. But it can be very difficult to convince oneself that what one is seeing on the Martian surface is actually a concave feature and not a convex one.

When examined carefully and from the correct perspective, the "worm" transforms itself into a canyon. No translucent tube, no ribs. Just steep canyon walls with sunlight illuminating one side. It is the inversion of the bright sunlit wall that gives the impression of a transparent glass tunnel through which the floor of the canyon is visible.

As for the transverse features giving the appearance of "ribs" on the "tube," they are most likely ripple-like sand dunes on the canyon floor or, alternatively, channels in the rock or some other perfectly natural feature.

If giant glass worms were not enough, what about little green men?

True. Well, maybe not *completely* true...

Early in 2009 an image beamed back from the Spirit Martian lander revealed a feature that superficially looked like a human figure sitting on one of the Martian rocks. This Martian manikin even had the canonical greenish tinge in the color image published on the NASA page!

Needless to say, a small number of websites published this as "proof" of Martian life. Either the manikin was a statue, even compared by some to the famous Little Mermaid of Copenhagen, or it was a real-life Martian taking a rest on the nearest comfortable rock.

Fortunately, most of the responses to these suggestions showed a healthy skepticism. One correspondent suggested that the entire thing was a hoax, an image of a seated person or figurine added to a genuine Martian panorama. This is apparently not the case, as the manikin is present in the original image published at the official site.

The "Manikin of Mars". NASA image.

Another correspondent pointed out that the figure is clearly quite close to the camera and must therefore be very small, maybe only a few inches tall. That is not very large, even for a Martian!

Of course, there is no real doubt that the manikin is anything more exotic than an oddly shaped bump on a slab of rock. Such formations are frequently seen on Earth, where they seldom arouse comment. But let one be photographed on Mars, and, we can be sure, somebody will hold it up as evidence for Martian life.

The Spiders of Mars

Mention of "spiders" in association with Mars might bring to mind David Bowie's *Ziggy Stardust and the Spiders from Mars*, but the spiders being referred to here are not actually *from* Mars but *on* Mars. And they are not really spiders either; they just look that way.

Like the "worm" and the true explanation of the Face, these features were discovered by the *Mars Global Surveyor*. They have only been found, thus far, at the south polar cap and appear as spidery or web-like patterns on top of the residual water ice – cap following the seasonal sublimation of the more volatile layer of frozen carbon dioxide. The "spiders" form round lobed structures and generally radiate from some central point, which is sometimes (thought not always) a crater.

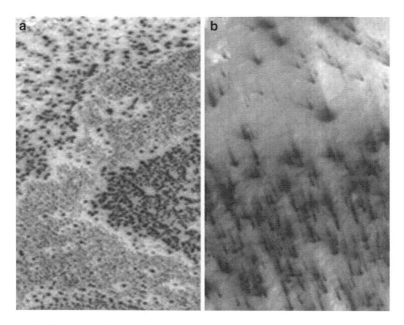

Spots and short streaks on the Martian ice. NASA image.

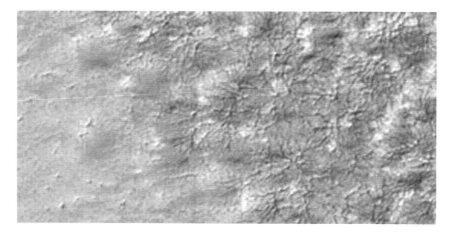

Spider-like formations appear with the Martian spring. NASA image.

The most likely explanation for these strange formations is as follows:

As spring warmth returns to the frigid polar cap, carbon dioxide ice lying beneath the ice cap's topmost layer bursts through and erupts in powerful geyser-like jets, blowing fine dark sand high into the air. Spreading outwards from the eruption, the dark sand eventually settles over the residual ice in the observed spidery patterns.

As Arizona State University's Phil Christensen explains, "The whole process begins during Mars' frigid Antarctic winter, when temperatures drop to –200°F. That's so cold that the Martian air – 95% carbon dioxide – freezes out directly onto the surface of the permanent polar cap.

"This seasonal deposit begins as a dusty layer of CO_2 frost. Over the winter, the frost recrystallizes and becomes denser … The dust and sand particles caught in the frost slowly sink. By spring, with the Sun about to rise, the frost layer has become a slab of semi-transparent ice about 3 ft thick, lying on a substrate of dark sand and dust.

"Sunlight passing through the slab reaches the dark material and warms it enough that the ice touching the ground sublimates … As days pass and the Sun rises higher, sublimation continues. Before long, the warmed substrate generates a reservoir of pressurized gas under the slab, lifting it off the ground.

"Soon after, weak spots in the slab break through, forming narrow vents, and high-pressure gas roars out at speeds of 100 miles per hour or more. Under the slab, the gas erodes the ground as it rushes toward the vents, snatching up loose particles of sand and carving networks of grooves that converge on the vents."

Once this becomes established, the vent will form in the same place year after year.

The weblike network resulting from these fantastic eruptions gives the impression of organic growth, and this initially raised speculation that the features might be colonies of simple organisms. However, the above explanation appears completely adequate to account for them and is in many ways even more interesting. Here we have a process, literally like nothing on Earth, helping to sculpture a region of our neighboring planet; in the meantime reminding us that not every amazing phenomenon of which nature is capable is represented on our home world. This alone should caution us as to how we interpret features found on other orbs in the cosmos!

Well, Then, Is There Really Life on Mars?

As a final word on Mars, leaving aside faces, ruined cities, glass worms, and other topics that are (putting it mildly) controversial, what, as we draw toward the close of the first decade of the twenty-first century, can sensibly be said about the prospects of finding life there?

The issue is in no small degree an emotional one. Let's face it. Most of us have more than strictly scientific curiosity about life on other planets. There is the romance of Neverland and Narnia – worlds of monsters and bizarre creatures, fairyland for adults, perhaps?

Of course, just because a belief has an emotional appeal has no bearing upon its truth or falsity, but it may subconsciously influence researchers to give more weight to evidence seemingly supportive of the belief. We have seen this influence many times in the life-on-Mars discussion.

On the other hand, there can also be an opposing tendency to become so aware of this danger that we become unwilling to admit evidence for life even if it were to bite us! Somehow, we have to strike the happy mean.

The happy mean is not easy to reach, of course, and any personal opinion will inevitably be biased to a greater or lesser degree. Nevertheless, certain facts do seem to be indisputable.

For a start, the concept of Martian life has shrunk in the past 110 years or so from that of intelligent creatures to plant life and then to micro-organisms, as our knowledge of the planet progressively unfolded. Mars might harbor the most Earth-like conditions in the Solar System, but it is still far from being a second Earth. In comparison with our home planet, Mars is a very harsh place indeed.

Secondly, there have been so many false alarms about Martian life that nobody should be too ready to propose it as an explanation for some Martian "anomaly."

In addition to the outright false alarms, there are also several "jury's still out" verdicts hanging fire. The arguments of Levin, Straat, and supporters – and the observations on which these arguments rest – have already been mentioned. Most authorities remain unconvinced by Levin's claim that *Viking* probably did find life on the planet, but the jury is still out. He may yet be vindicated.

Then there was the discovery of organic material and microscopic structures suggestive of fossilized bacteria in some meteorites whose origin was almost certainly the Red Planet. Back in the mid 1990s, news broadcasts carrying this story announced it as final proof that life had at least existed on Mars in the past, even if it might now be extinct. But the real final proof remained elusive, and there is still no consensus on what the meteorite organic traces really did or did not imply.

The same may be said for the discovery in 2003 of methane in Mars' atmosphere by three groups of Earth-bound astronomers. As remarked earlier, a similar claim in 1969 turned out to be another false alarm, but this later one is secure enough in broad terms. Methane is not stable for long periods in the atmosphere of Mars and dissociates relatively quickly due to the action of ultraviolet light. Its presence therefore implies a contemporary (or at least

A Martian meteorite. Organic structures in this, and similar, Martian rocks reaching Earth reopened the life debate in the 1990s. NASA image.

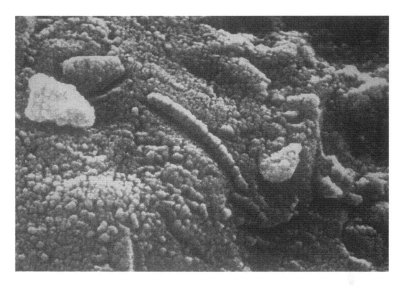

Structure found in a Martian meteorite. Despite its life-like appearance, the balance of opinion favors a non-biological explanation. NASA image.

very recent) source, with biological agents such as methanogenic bacteria high on the list of possible suspects.

An early objection to a biological explanation rested on the fact that the total amount of methane detected was extremely small – just 10.5 parts per billion – as well as being restricted to just a few places on the surface of the planet, principally the northern hemisphere regions Arabia Terra, Nili Fossae, and the southeastern quadrant of Syrtis Major. If the study of the biology of our planet has taught us anything, it is that life is adaptable. Presumably, this property is not restricted to life on Earth, unless life *itself* is restricted to Earth. If there really is life elsewhere, we have no reason for supposing it to be any less adaptable.

PROJECT 12
Syrtis Major

One of the regions on Mars where methane has been detected is arguably the most conspicuous feature on the planet. During oppositions of Mars, it is readily detected in telescopes as small as 3-in.

(7.5-cm), or even 2.5-in. (6-cm) refractors. When Mars is close to opposition and rising as the Sun sets, wait for several hours until the planet is high in the sky and then try viewing it in a 3-in. telescope at a magnification of about 100×. One of the polar caps (the one facing Earth at the time) should be visible and, if the right hemisphere is turned toward Earth, so should dark equatorial regions, including a rather conspicuous triangular wedge-shaped formation. This is Syrtis Major, and if it is not visible when you look, sooner or later it will come into view as Mars turns on its axis.

The only exception to this is during the occurrence of a global Martian dust storm. As luck would have it, these are more likely to occur when oppositions of the Red Planet occur around the time it traverses that part of its orbit nearest the Sun. As these "perihelic oppositions" also mean that Mars is then closest to Earth and at its largest apparent size, they can be infuriating for observers!

Assuming that there is no Martian dust storm, Syrtis Major should be an easy target. True, there will be no visible indication of methane emissions, but just seeing one of the sources of this gas from your own backyard is interesting!

Only a few decades ago, most astronomers mistook Syrtis Major for an especially vegetated area. Ironically, with the discovery of methane, it has once more entered into the enduring life-on-Mars debate.

In other words, if methanogenic bacteria, or any other living organisms, exist on Mars at all, we would expect to find them essentially covering the planet instead of being restricted to a mere handful of isolated oases. Mars should be well endowed with methanogens with a correspondingly higher concentration of methane in the planet's atmosphere. We will return to this point, and a possible "escape clause" from the objection, shortly.

Another possible source of trace methane is localized, low level, volcanic activity. This may seem to gain some superficial support from the identification of one of the methane sources as the southeastern part of Syrtis Major, a very conspicuous dark region once thought to be a Martian ocean and, later, a vast field of vegetation. It is now identified as an ancient volcano, and given the methane emission, it might be wondered if it is altogether extinct!

However, if volcanism – even at a very low level – really is the culprit, other volcanic gases such as sulfur dioxide should also be present together with the methane. Yet, no such gases have as yet been detected.

A more or less distantly related process that occurs in a few places on Earth is serpentinization. This is a water/rock reaction that can occur when major fracturing and faulting exposes mantle-like minerals to sea water or ground water. In the process, iron oxide is transformed into serpentine and methane.

Nevertheless, geologist Professor Lisa Pratt doubts that this process can account for Mars's methane. There is simply no evidence that the sort of deep faulting and uplift required for serpentinization actually takes place on Mars.

Alternatively, fragments of a comet relatively recently striking Mars and embedding themselves into the desert dust could add methane and all manner of organic traces to the atmosphere as they slowly evaporate away. This might seem to draw the long bow, but as it surely happens from time to time, it is not too outrageous to think of it having happened in *our* time.

On the other hand, though, it does seem to be pushing speculation to suggest *three* recent cometary impacts, although a single object disrupted into three fragments might be a possibility. Nevertheless, as methane is only a minor constituent of comets, the non-detection of other more prevalent gases (such as carbon monoxide and cyanogen) makes this explanation very doubtful.

Yet, unless some even more exotic explanation for the methane has been overlooked, one of these suggestions *has* to be right!

Early in 2009, more detailed results from the ongoing study of Martian methane detections may have removed some of the objections to a biological source. For one thing, even though the overall percentage of methane in the Martian atmosphere is minuscule, the concentration over the three detected source regions, although not large, is not entirely to be sniffed at (pardon the pun). According to study author, Michael Mumma, the primary plume at one time contained 21,000 tons (19,000 m.t.) of methane. The gas was being belched out at the rate of about 1.3 pounds (0.6 kg) each second.

Yet, outside of the plumes, the methane concentration is extremely low. So low in fact, that something other than solar UV

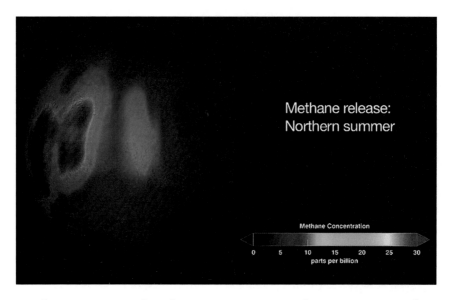

Map showing regions of methane concentration in the Martian atmosphere. NASA image.

must be breaking it down! The methane destroyer, Mumma and colleagues surmise, is an oxidizing agent or mixture of oxidizing agents in the planet's soil. We know that Martian winds continually loft dust high into the air. Dust storms are common, and whirlwinds are forever marching across the planet's surface. Some of these rotating columns of wind and dust reach 5 miles in height. Carried into the air as coatings on dust particles, the methane-destroying chemical(s) probably account for the unusually rapid disappearance of this gas and explain its very low average concentration in the Martian atmosphere.

If that is true, the objection that the gas cannot be biogenic because it is simply too sparse may be, at least in part, countered. Indeed, if the Martian soil itself is destroying methane, the amount actually being generated may be much greater than even the quantity detected in the plumes suggests, especially if the source lies well beneath the surface of the planet.

The plumes appeared to be most active in the late Martian summer, but it is not yet known whether they keep recurring in the same places or, if they do, how frequent this occurrence is. The methane that has been detected may have escaped with a

Dust devils on Mars. Chemicals in dust lifted into the Martian atmosphere by whirlwinds and dust storms may help to reduce the global atmospheric methane content on Mars. NASA image.

melting of subsurface ice, releasing pockets of gas that had accumulated since the previous breakout. If that is true, the sources identified might not be isolated regions where the gas is generated so much as isolated regions where conditions simply favor its partial escape into the atmosphere. This also tends to weaken the objection to a biological origin in so far as isolated methane emissions may no longer *necessarily* imply that the hypothetical organisms are themselves confined to pockets or oases.

As mentioned earlier, most scientists are justifiably skeptical about oases of life existing on Mars, i.e., life confined to isolated spots on an otherwise sterile planet. If it is there at all, the wider feeling is that it should be everywhere, or very nearly everywhere. This is, admittedly, a prejudiced view based on our experience with life on Earth, but as Earth life is the only example we have at present, the prejudice is understandable and probably justifiable.

The same is true to a lesser degree of "oases" in time or, to put it another way, the hypothesis that once upon a time Mars was teeming with life, but that all has long since become extinct. Again, from the experience of our own planet, life adapts over changing conditions of time pretty much as efficiently as it does over changing conditions across the face of the planet. In fact, the two tend to go together. Species become extinct, but life per se goes on.

If that is true, it implies that if there ever was life on Mars, it is still there, in some form, today. Conversely, if contemporary Mars is sterile, it has probably always been that way.

Nevertheless, it might be argued that a lifeless Mars would be rather surprising. As exploration of the planet continues, the harsh ideas of the 1960s and 1970s have been modified somewhat. Water (in the form of ice) has been found, and there is evidence that Mars once supported large bodies of liquid water on its surface. Even today, flows of liquid water probably exist for short periods and droplets may persist in the soil. In other words, Mars now looks somewhat less hostile than it did circa 1970, and there is good reason to think that this planet and our own dear Earth harbored not-too-dissimilar conditions in the distant past – about the time life first appeared here. Without modifying our earlier statement that life is the most complex and therefore the most improbable of phenomena, we must nevertheless not forget that this improbable phenomenon took root on at least one planet – our own. If it did not appear on Mars as well, under conditions that were then rather similar, that fact itself would require an explanation.

Moreover, even if life did not experience a second genesis on Mars, rocks blasted from Earth during the pounding that our planet received from asteroids in its youth surely reached our neighbor. As there are hints that microbial life existed on Earth towards the close of this period, some of these space-faring rocks may have held dormant microbes. In any case, although giant impacts became rare (fortunately for us!), after the cessation of the early bombardment, they did not cease altogether after life became well established. From what we know about the durability of microbes in their dormant state, some of the microorganisms blasted out by these impacts should still have been viable when they reached Mars. Awaking in an environment not too different from the one in which they were ejected, they would settle right into their new home and start reproducing!

If it is true that once life – whether indigenous or introduced – gets going on a planet it should (excepting some tremendous change of conditions) still be present, the question is "Where is it on Mars?" "Why has it not been unambiguously detected there?"

Even if it is only microscopic, it is unlikely to be completely concealed from view. Many microscopic organisms make a macroscopic structure, as the stromatolites of western Australia testify. Also the coral polyp, though not microscopic, is surely tiny but is still responsible for creating structures (coral reefs) so large that some are visible from Earth orbit.

It is possible, though perhaps unlikely, that analogous features on Mars have been seen and not recognized. Perhaps we are missing evidence of Martian life because we can't see the forest for the trees. The discovery of rock varnish, or what appears to be rock varnish, was thought by some to have been such, but the significance of this remains controversial. Maybe – to pick an example at random – the red soil of Mars is indicative of the work of iron-oxidizing bacteria (unlikely, but the suggestion has been made). In that case, *every* image of Mars includes signs of life, but signs too subtle to be easily recognized!

Alternatively, it might be suggested that even if there is no life *on* Mars, there may still be life *in* Mars.

This is not as question-begging as it sounds, and let us stress that we are not talking about Martian gnomes!

We are talking about microscopic life forms similar to Earth's *Bacillus infernus.* The name literally means "the bacillus from hell," but this is not meant to imply that it is an evil little bug. On the contrary, it refers to the hellish environment in which the bacillus thrives, miles beneath our feet!

This organism is about as far removed from the rest of terrestrial life as it can possibly be. It does not rely on the Sun, doesn't use photosynthesis, and doesn't consume organic material synthesized by other organisms. It lives on and in rock and has a metabolic rate that makes a sloth look like a speed freak. It takes life so easy that it only divides about once in a millennium!

If Earth were flung into interstellar space (such that the atmosphere froze solid on its surface) or took up an orbit inside that of Mercury (and the oceans boiled away and rocks glowed red with the heat), *B. infernus* would continue as if nothing had happened.

Could something akin to *B. infernus* exist deep within Mars, whether or not anything lives on the surface of that planet?

Who can say? But if something like that does thrive there, it will not be easy to find without some very deep drilling.

Whatever the truth of the matter, we can only hope that space exploration will settle the question one way or another in the not-too-distant future. Whichever way the chips finally fall, the solution to this long mystery will certainly be interesting.

Weird Venus

Mars is not the only planet to arouse our wonder and curiosity. Even more mysterious is the brilliant Venus, physically far more Earthlike than Mars, yet literally shrouded in a cloak of mystery that has only been lifted with the advent of space exploration.

So beautiful from afar, we now know that Venus is truly a dreadful place. Furnace hot and with a crushing atmosphere, surrounded by a thick haze of concentrated sulfuric acid, the place sounds not unlike a literalist vision of a fire and brimstone hell. Except that there is no actual fire. The atmosphere cannot even support that!

This appreciation of our closest planetary neighbor has come, however, only slowly, and it was not too many decades ago that a Venus covered by watery swamps or even oceans seemed a likely possibility. On such a world as that, life, we thought, might well have been flourishing.

Nevertheless, life was not assumed as an explanation for anomalous observations as readily as it was for Mars. The very first astronomers to peer through a telescope at the planet may have expected it to reveal clear surface markings, and, had that happened, no doubt speculation about who or what might have inhabited this world would have become as rife as it did with Mars. But telescopic views revealed little; just a brilliantly white orb that (because of its position sunward of Earth) went through phases similar to those of the Moon.

Or, *almost* similar to those of the Moon! The cusps of the planet in crescent phase were often seen as greatly extended, and from this it was correctly inferred that unlike the Moon, Venus possessed a dense atmosphere. This explained the almost total lack of visible surface detail. The atmosphere was an exceptionally cloudy one, though for a long while the nature of those clouds (water? dust? something else?) remained the subject of debate.

The picture that eventually emerged was a far from pretty one. Much of the "cloud" layer is not so much cloud in the true sense of the word as a dense pall of sulfuric acid "smog." Such is our neighboring world!

Canals on Venus?

Just about everyone has heard of Percival Lowell's controversial observations of "canals" on Mars, but not as many are aware that he also observed long straight lines on Venus. Needless to say, these observations were very controversial and drew quite a deal of criticism from other astronomers for whom the planet remained merely a brilliant but featureless orb.

Using the 24-in. (60-cm) Clark refractor at Flagstaff in October 1896, Lowell detected a series of markings on Venus that he described to the Boston Scientific Society as being "surprisingly distinct; in the matter of contrast as accentuated, in good seeing, as the markings on the Moon and, owing to their character, much easier to draw." According to Lowell, these markings took the form of long finger-like streaks "which started from the planet's periphery and ran inwards to a point not very distant from the center … well-defined and broad at the edge, dwindling and growing fainter as they proceeded, requiring the best of definition for their following to the central hub." Superficially similar to the canals on Mars, Lowell nevertheless stressed that "there is about them nothing of the artificiality observable in the lines of Mars. They have the look of being purely natural."

Lowell's report was not received well by the majority of astronomers. Although there was nothing very radical about the claim itself, the mere fact of seeing markings on Venus was so at odds with just about everyone else's experiences that the claim was made to sound outrageous.

Perhaps the streaks were simply optical illusions induced by staring at the brilliant face of Venus. Or maybe something was wrong with the telescope, a suggestion made by Lick Observatory director, Edward S. Holden. According to Holden, the streaks might have been caused by a strain on the telescope's object glass "induced by an overtight condition of the adjusting screws or of the objective in the cell."

This last remark drew a hasty and predictably sharp response from the telescope's maker Alvan G. Clark. Clark, who had made the large refractor at Lick as well as the one at Flagstaff, retorted that "I personally superintended the adjustment of the Lowell objective in its cell at Flagstaff before the observations in question were made" and "the same class of strain which exists in the Lowell [i.e. Flagstaff] must be present also in the Lick objective."

Lack of confirmation of the features apparently caused Lowell to doubt his own eyes for a while. With no similar reports from other observers, by 1902 he seems to have been on the brink of dismissing his results, although this attitude changed again over the next five years, and for the rest of his life he expressed no doubt as to the validity of what he saw. Moreover, he used his observations to draw some unconventional conclusions about the nature of the planet.

Because the markings seemed to occupy the same place at every observation, he concluded that they must be surface features. That, in turn, implied that the atmosphere (although indisputably dense) was not opaque as widely believed but instead quite clear and transparent. What was seen through telescopes was not the top of a cloud layer but the true planetary surface.

Lowell also came to accept the opinion of G. V. Schiaparelli that Venus was in a 1:1 spin orbit lock with its axis of rotation perpendicular to its orbital plane. This fixed, Lowell concluded, the position of the terminator.

Lowell's model of Venus, as detailed in his 1909 work called *The Evolution of Worlds*, was of a world radically divided between eternal day and everlasting night. One side of the planet was perpetually scorched by sunlight, while the other lay in perpetual deep freeze and unbroken darkness. He argued that rising air over the hot side produces something vaguely analogous to the sea breezes of Earth, except that the resultant inflow of cold air could hardly be described as a "breeze." The inflows would be, in Lowell's scenario, "indraughts ... of tremendous power [as] a funnel-like rise [at] the center of the illuminated hemisphere, and the partial vacuum thus formed would be filled by air drawn from the periphery which, in its turn, would draw from the regions of the night side."

It was this effect that, he reasoned, caused the streaks. He wrote that "Such winds would sweep the surface as they entered, becoming less superficial as they advanced, and the marks of their inrush might well be discernible even at the distance [of Earth].

Deltas of such inroad would thus seam the bounding circle of light and shade."

According to Lowell, Venus was a harsh, sterile, and desiccated world. His belief in a living Mars was clearly not matched by similar opinions about Venus. As we earlier saw in connection with his skepticism concerning alleged signals received from Mars, Lowell was not someone who readily concluded in favor of extraterrestrial life as an explanation for anomalies on other worlds. This is a point worth remembering in view of his somewhat dubious Martian legacy.

Ironically, the year Lowell died (1916) saw two apparent confirmations of his observations by astronomers Maxwell and Wilson. The latter made a similar observation the following year as well. Following this, another report came from F. Seagrave in 1919, and there were several others during the 1930s.

Then, in the 1950s and early 1960s, French amateur astronomer Ch. Boyer made ultraviolet observations of the planet, the results of which led to the discovery of the 4-day superrotation of Venus's upper atmosphere. The visible sign of this is a streaklike circulation pattern. Moreover, studies of wind velocities in the upper atmosphere of the planet indicate that this superrotation pattern gives way, at about 62 miles (100 km) altitude, to a subsolar and antisolar flow not unlike that suggested by Lowell. So maybe Lowell was quite correct in what he saw and partially correct in his interpretation, even though his insistence that he was seeing surface features was incorrect and his model of Venus did not stand the test of time.

The markings reported by Lowell are not the only features that have been reported on this traditionally featureless planet. On a number of occasions, white spots were also seen by telescopic observers. Typically, these were reported near the poles of the planet, but unless the controversial clear atmosphere model of Lowell was correct, polar ice caps a la Mars and Earth could not be the explanation. Early observers suggested that the spots might be the summits of mountains poking above the cloud. Some astronomers thought that they were seeing peaks over five times the height of Mt. Everest!

The polar spots have now been identified as the cloud swirls first imaged by *Pioneer Venus*. Similar spots occasionally reported near the terminator but at much lower Venusian latitude are undoubtedly isolated cloud cells.

The Strange Glow of the Venusian Night

Venus goes through phases like our own Moon, though on a different time cycle, of course. At so-called inferior conjunction, when the planet is more or less between Earth and the Sun and therefore also near its closest approach to us, it enters the "new" phase. Immediately prior to, and following, this phase it is visible in telescopes as a thin crescent.

We all know that when the Moon is in the crescent phase, it does not appear simply as a bright crescent. We also see a faint outline of the entire lunar disk sometimes picturesquely referred to as the "old Moon in the new Moon's arms." This is caused by earthshine reflecting back to us from the Moon's darkened hemisphere and, as we said in Chapter One, was an early indication that Earth is truly a planet.

Oddly, what looks like the same phenomenon has been reported many times during the crescent phase of Venus!

As long ago as 1643, J. B. Riccioli, a Jesuit professor of astronomy at Bologna, suspected that he could trace the entire disk of the planet. The phenomenon was definitely seen by British clergyman and astronomer William Derham in 1715.

This presents something of a mystery. For an observer on the Moon, our world shines with a brilliance some 80 times greater than the full Moon appears to us. This is ample to light up the lunar night and reflect its share back to us. But even though hypothetical observers above the clouds of Venus would see Earth gleaming brighter than Venus ever appears to us, its light is hardly sufficient to illuminate the cloudscape enough for it to be visible back on Earth. Something else must be happening there.

Over the years, any numbers of suggestions have been made to explain the "Ashen Light," as it is officially known.

Early proposals included some involving life of one form or another; rotting vegetation, oceans filled with phosphorescent organisms, even artificial lighting set by intelligent beings. The wildest suggestion was probably that proposed in 1840 by Franz von Gruithuisen, who thought the glow might be from festival lights lit to celebrate the crowning of a new Venusian ruler! (As an aside, it should be noted that this astronomer was also the first to suggest the far more plausible, and now generally accepted, theory that Moon craters were caused by impacting meteorites. Sometimes bizarre and plausible ideas can issue from the same person!)

A fictional suggestion (and we stress that it *was* fictional) even had the clouds of Venus inhabited by living lights.

At the other end of the spectrum, the Light was dismissed by some as nothing more than illusion, possibly inspired by the subconscious association of the crescent Venus with the crescent Moon. The mind simply filled in the missing earthshine.

In between these extremes were suggestions of auroral activity, lightning (or perhaps the St. Elmo's fire type of discharge sometimes seen over the tops of terrestrial thunder clouds), or the refraction of sunlight through the dense atmosphere.

One plausible hypothesis suggested something akin to airglow as the culprit. Exposed to strong solar ultraviolet light, carbon dioxide molecules in the planet's upper atmosphere are split into carbon monoxide and single atoms of oxygen. When two of the latter combine to form molecular oxygen, a flash of green light is emitted, and the combined effect of these flashes should cause the upper atmosphere of Venus to softly glow. This surely occurs, but the resulting glow is not strong enough to add any significant contribution to the Ashen Light. A far stronger source of illumination is needed to explain the observations.

Another interesting proposal was made by Lowell's friend Godfrey Sykes and accepted by Lowell on the basis of his model of a Venus having a transparent atmosphere. Lowell, as we have seen, pictured the night side of Venus as a land of perpetual cold. (Apparently, he did not consider atmospheric heat transfer sufficient to raise the temperature significantly.) This hemisphere was, in his opinion, covered with eternal ice, and on this basis Sykes suggested that the Ashen Light might be a "reflection from the ice-fields there of light received from the Earth, the other planets and the stars."

That explanation sinks with the clear-atmosphere model of Venus, although there have been informal suggestions that Venusian clouds themselves might reflect enough Earth light to give rise to the illumination. As already remarked, however, Earth is simply not bright enough (as seen from Venus) to account for the degree of illumination observed.

It was not until infrared images of Venus were obtained that a clue to the real nature of the Ashen Light was found. In the IR, even the night side of the planet is brightly luminous. It is as if the Ashen Light shines brighter in the IR, which is understandable given the conditions now known to exist at the planet's surface.

With a surface atmospheric pressure 100 times greater than Earth's and "air" composed mostly of carbon dioxide, Venus experiences a horrendous planetary greenhouse. Night and day, the planet's surface is roasting hot, and it is the radiated heat from its scorching rocky surface that gives rise to its IR luminosity. But with rocks glowing red hot, not all the radiation is in the IR wavelength. Half a planet worth of red-hot rocks would send up quite a glow. Enough to be seen from Earth as the Ashen Light?

If these thoughts are correct, this soft-glowing luminescence becomes yet another message from the planet telling us just how hostile a place it really is.

PROJECT 13
Seeing the (Ashen) Light*

The Ashen Light would undoubtedly be far easier to observe were it not for the dazzlingly bright presence of the crescent Venus. There may be an optimum time between dark sky and daylight when the glare of the planet is somewhat reduced while the faint glow of the Ashen Light is not totally overpowered by twilight. Careful monitoring of the planet in a lightening dawn sky may prove successful. At least, the experiment seems worth trying for a serious Ashen Light hunter.

Some visual astronomers have sought the light with an occulting bar, especially one shaped like a crescent to conceal the similarly shaped bright side of Venus. In theory, this device blots out the brilliant crescent while allowing the nighttime side of the planet to remain unobscured. In practice, keeping Venus at the right position with respect to the bar is not easy, and the planet's light scattered in our atmosphere remains a problem even with the crescent hidden, but some successes have apparently been reported using this method.

Another possibility is a brief view of the light during those times when the crescent Venus is occulted by the dark region of the crescent Moon. For the serious Ashen Light hunter, these rare opportunities are not to be missed, and a quality video of the Ashen-Light-illuminated dark side of Venus sinking behind the earthshine-lit portion of the Moon would certainly be a prized possession for any observer!

A Satellite of Venus?

Earth has its Moon, Mars its duet of Demos and Phobos, but Venus, like Mercury, is without a companion. Although it is just possible that one or more tiny meteoric satellites orbit Venus (and maybe Earth as well), there is as yet nothing to support this conjecture, and, in any case, it is absolutely certain that no sizable body orbits our closest planetary neighbor.

Yet, on a number of occasions, several reputable astronomers apparently saw a small bright object close to Venus.

Thus G. D. Cassini, to name one famous early observer, wrote in his journal for August 28, 1686, that "At 4.15 a.m. while examining Venus with a telescope of 34 ft focal length, I saw at 3/5 of its diameter to the east an ill-defined light, which seemed to imitate the phase of Venus, but its western edge was more flattened."

In the years following, especially between 1761 and 1764, sightings of a bright object very close to Venus multiplied. There were at least 33 reports by 15 different observers during the seventeenth and eighteenth centuries, 18 in the peak year of 1761 alone. Yet, with one interesting exception that we shall look at shortly, sightings ceased after 1768.

Was there once a Venusian moon that for some reason met disaster before 1768?

Highly improbable, to say the least!

For a start, observations of a large object showing distinct phases (such as Cassini saw) can quite easily be explained as ghost reflections of the planet itself, an explanation put forward and closely argued as early as 1766 by priest and astronomer Father Hell (a somewhat unfortunate name for a priest, one would think!). By the way, the Moon crater named Hell is named for this gentleman and is no reflection of the conditions prevailing there.

Other purported satellite sightings have been identified with stars, and one – by Roedkiaer at Copenhagen on March 5, 1761 – may have been of the planet Uranus, some 20 years prior to its official discovery by Herschel.

One, however, is definitely strange. This is the sighting of a bright star-like object near Venus made by no less an astronomer than E. E. Barnard on the morning of August 13, 1892.

Writing in *Astronomische Nachrichten* some 14 years later, Barnard recalls that "While examining Venus ... with the 36-in. of the Lick Observatory ... I saw a star in the field with the planet. This star was estimated to be at least the seventh magnitude. The position was so low that it was necessary to stand upon the high railing of a tall observing chair. It was not possible to make any measures, as I had to hold on to the telescope with both hands to keep from falling. The star was estimated to be 1' south of Venus and 14" preceding." He then gives the estimated position, which does not correspond with any star of that brightness. The observation was made in very bright twilight, just half an hour prior to sunrise.

Barnard stresses that there was no doubt in his mind as to the validity of the observation, and he completely dismisses the possibility of an optical ghost. He also notes that the brighter asteroids were not near the position at that time.

Barnard did not associate this with a satellite of Venus, but he did offer that "it does not preclude the possibility of its being a planet interior to Venus," though straightway adding that "such is not probable." He also dismissed the possibility of Vulcan, as the elongation of Venus was 38° at the time, greater than an intra-Mercurial planet could attain, though not too great for one orbiting *between* Mercury and Venus.

This observation remains a mystery. The nucleus of a comet might possibly explain the observation, although one would think that some sign of the fuzzy coma would have been noticed at that elongation.

Another, and perhaps more promising, possibility is a very fast nova. Considering that the position given by Barnard placed the object within the Milky Way band, where most bright novae occur, this explanation actually has quite a lot in its favor. It would be interesting to know if searches were made for the object on subsequent mornings, but Barnard's account does not detail this.

But whatever it was that Barnard saw, it was certainly neither a satellite of Venus, nor an intra-Venusian planet!

The Venus Strain?

As we will see later in this book, alien organisms have at times been suspected of washing to Earth in rainfall. A recent example that caused quite a stir was the red rain falling in India, about which more will be said soon. But an even more puzzling story concerns some very ordinary transparent rain falling on the Norman Lockyer Observatory in Sidmouth, England, which, some say, may have contained micro-organisms from Venus!

The story goes like this. Long time observatory director, Donald Barber, exposed many photographic plates recording variable star spectra between the years 1936 and 1963, which he developed at the laboratory using the observatory's supply of fresh (unchlorinated) water. Most of the plates were developed without trouble, but Barber noted an annoying tendency for sporadic failures to occur. These were identified as attacks by a kind of water-borne yeast-like bacteria that liquefied the gelatin emulsion and appeared unusually immune to both silver and silver halides. Nine of these bacterial infestations occurred between the above-mentioned years, and further investigation found that another event probably occurred in 1930 and yet another may have taken place two years later.

Samples of the microorganisms were taken to the Lister Institute, but the organism was never identified as a known species. Then Barber noted something really weird.

In nearly every instance, the infestation followed soon after an inferior conjunction of Venus at a time when this was coupled with a strong geomagnetic storm. In the six best determined events, the time of the infestation averaged 55 days after the conjunction, the shortest interval being 35 days and the longest 67. There also appeared to be a correlation with northerly wind patterns and heavy rain preceding the beginning of the infestation.

In the light of this, Barber put forward the following radical suggestion. The microorganisms responsible for this damage to the plates came from Venus!

By comparison with our home planet, Venus has a negligible magnetic field. This means that the solar wind (the stream of protons constantly boiling off the solar corona) continually strikes its outer atmosphere and can potentially repel small particles into space, carrying them off in a direction away from the Sun. Particles as small as bacteria could, theoretically, be swept out of the upper atmosphere of Venus and transported to Earth on this solar wind.

From what we have learned over the years about bacteria, there is little doubt that at least some of these microbial astronauts would still be viable upon arrival at Earth and could become active once they found themselves again in a planetary environment. They would then do what bacteria do best and colonize their surroundings.

At first sight, Venus surely seems the last place we might expect to find life – even bacterial life. Certainly, the planet's surface must be sterile. We cannot conceive of even the most extreme of extremophiles surviving on a world where the rocks glow red hot.

But the upper atmosphere might be a different matter. It is possible that bacterial colonies *could* thrive there. Most planetary scientists think that in the early years of the Solar System, Venus was a much more benign place, quite possibly with bodies of water existing on its surface. If so, primitive life may have taken hold on the planet about the same time that the first microorganisms appeared on Earth. Over millions of years, as the planet heated up and all bodies of water eventually boiled away, most of this life would have been destroyed, but any organisms that found their ecological niche in the upper atmosphere could have adapted well to the changing environment and might still be there today.

Alternatively, we can be pretty sure that any rocks blasted from the early Earth during the Late Heavy Bombardment arrived on Venus as meteorites, just as Martian rocks have reached Earth in more recent times. As there is some evidence (albeit, not universally agreed upon) that early life existed on our planet toward the end of that period, it is not inconceivable that terrestrial life was transferred to Venus in those early days and that the remote descendents of this life still inhabits the Venusian upper atmosphere. (We recall that a similar argument was made for the transportation of early terrestrial microorganisms to Mars. Actually, calculations of the respective amount of terrestrial material reaching either planet show that Venus would have received more Earth rocks than Mars, so transportation of early organisms to Venus appears quite probable. Their subsequent history may, however, be another matter).

Although the suggestion of microorganisms in the upper Venusian atmosphere is based on very slender evidence, there may actually be some hints in its favor. Analysis of light reflected by the uppermost haze level of Venus indicates that this is largely comprised of sulfur-coated particles of an unknown nature, whose size and shape are at least consistent with bacteria. This does not necessarily mean that they are bacteria, of course, just that (at minimum) observations have not ruled out that possibility. Actually, as a number of people have pointed out, a coating of sulfur (especially in the S_8 form) would be very beneficial to bacteria high in the Venusian atmosphere, as this would provide them with a very efficient shield against the powerful ultraviolet rays of the Sun.

Back to the Norman Lockyer bacterial invasions, Barber's suggested scenario ran like this:

Powerful disturbances in the solar wind (of the type causing magnetic storms on Earth) expel large quantities of bacteria from the outer regions of Venus' atmosphere. These get swept into long plumes in the ante-solar direction from the planet and eventually reach beyond the orbit of Earth. When these solar wind events coincide with inferior conjunctions of Venus, Earth comes into a position where it can pass through the bacterial plume, picking up some of its content. These bacterial particles are then funneled down to the polar regions by our planet's magnetic field and enter into the general atmospheric circulation. In the Norman Lockyer case, Barber envisioned them being carried southward by northerly winds, incorporated into rain clouds, and from thence to the observatory's water storage.

As a general scenario, nothing about this is impossible, but it does run into some specific problems which must leave us skeptical. For a start, there is no clear link between the bugs at Norman

Lockyer and the sulfur-coated particles suspended in the air of Venus. Nothing reported suggested that the bacteria isolated in the observatory's water were coated with this element or that they displayed any obvious sign of adaptation to a highly sulfurous, strongly UV irradiated environment.

Likewise, any bug from Venus would, presumably, have sulfur high on its menu. But why would it consume gelatin? Or be tolerant of silver and silver salts? The answers are far from obvious.

Furthermore, there was, apparently, nothing about the microorganisms that suggested a non-terrestrial origin. There is no suggestion that they differed significantly from any other bacterial species. Of course, we know so little about the parameters of life that this cannot be the last word on the subject. Maybe all life anywhere in the universe has the same properties. Moreover, if life was transported from Earth to Venus via early terrestrial meteorites (or, conceivably, from Venus to Earth via early solar wind), it would be *expected* to show basic similarities.

Lastly, it is strange that an extraterrestrial strain of bacteria should only show up (and repeatedly at that!) in one place. Barber's explanation that most observatories use chlorinated town water cannot be the whole story. Surely, somebody else must have been troubled by the bacteria!

So, what are we to say about this?

No firm conclusions. The coincidence is certainly interesting, but it would have been more convincing had the bacteria been sulfur-coated, sulfur-metabolizing thermophiles. This, they apparently were not. Yet, nothing that has been said positively rules out a Venusian origin either. All that we can do is keep a weather eye on the rain!

Which Is the Really Weird World: Venus or Earth?

The search is on for terrestrial planets in orbit around other stars, and there is naturally a good deal of interest in this quest. After all, while finding an analog of Jupiter spinning around a distant Sun might be interesting, there is something deeply appealing about finding a second Earth.

Yet, it is easy to get too carried away with enthusiasm. The news media – even serious scientific news media – all too often blur the ambiguity inherent in the precise meaning of "Earth-like." Does this mean a planet physically like our own (similar mass, size, density, and so forth) or does it mean one on which we could make our home and (just maybe) on which folk not too different from us already have?

If we mean the former, an Earth-like planet has already been found; not around a distant star, but right next door in our own Solar System. In fact, it is the closest planet of them all – Venus. By all physical measurements, Venus is a terrestrial planet. True, it is not an exact clone of Earth (a little smaller, a little less massive, and slightly less dense), but it is near enough to be thought of as Earth's twin and is by far the most (physically) Earth-like planet in the Sun's family.

And yet, of all the small rocky bodies in the Solar System, it is undoubtedly the least inviting to us. Even a visit to Mercury sounds like a better deal! In this respect, Venus is about as different from Earth as a "terrestrial" planet could be.

So we have an odd situation. Two terrestrial planets of similar size and mass. One is benign and habitable. The other, a scorching world crushed by a horrendous atmosphere and swathed in acidic clouds.

The question is, "Which (if either) is typical of terrestrial planets?" Or, perhaps we could ask "Should we really call planets of this size and mass 'terrestrial' or should we really be calling them 'Venusian'?"

In earlier times, Earth chauvinism (though disguised as "Copernican considerations") assumed that Earth was the benchmark for planets. At first, Venus was assumed to be rather Earth-like, and when this illusion was shattered by the advent of robotic space probes, the question implicitly asked was, "What went wrong with Venus?" If Venus failed to turn out to be another Earth, something must have gone amiss with its evolution.

But as we learned more about our own and neighboring worlds, a surprising twist to this question emerged. Nothing went "wrong" with the evolution of Venus at all. Venus appears to have evolved just as a planet of this variety should. It was the evolution of Earth that "went wrong"! Something happened to our own dear planet that thwarted its "natural" growth into a second Venus. Something pushed it in a weird direction off the evolutionary track. And we are the result!

That "something" was the giant impact that formed the Moon. We have already seen how the presence of the Moon stabilizes Earth's axis and, as a result, its climate, in addition to regulating the length of our day.

But the giant impact itself had some very significant effects on our planet quite apart from the existence of the Moon per se.

For instance, had the impact not removed much of our planet's crust (some of which became incorporated into the Moon itself), plate tectonics would have seized up long ago. This is what happened to Venus. Having experienced no Moon-forming impact, the crust of Venus is some four times thicker than ours. By contrast and thanks to the impact, Earth's crust is thin and broken into segments which can slide over each other in a process that is not only responsible for mountain building but is also widely thought to be vital to the existence of complex life on this planet. Thanks to plate tectonics, the level of carbon dioxide in our atmosphere is regulated (conveniently keeping the planetary thermostat at an even keel) and nutrients are distributed as the crustal plates move. Moreover, because the tectonic process is a sort of "engine" that uses energy in its "work," heat from the planetary core is dissipated steadily and constantly. If plate tectonics seized up or never got going in the first place, internal heat from radioactive decay would just keep building under the planet's thick and immovable crust until, like a boiler whose safety valve has been soldered shut, pressure reaches a critical level and ... BANG!

"BANG" in this context means global volcanism of such ferocity as we can hardly imagine. This happened on Venus millions of years ago, flooding the entire planet with an ocean of molten lava and leaving its continuing legacy in the dense atmosphere and sulfurous clouds that give the planet its unappealing character.

This is what "should" have happened to Earth. Fortunately for us, a Mars-sized planet got in the road, gave us the Moon, plate tectonics, a thinner atmosphere, a stable axis of rotation, and a planet that we can call home!

So, really, the issue as to whether Venus or Earth is more typical comes down to the frequency of similar giant impacts as planets are being formed. In 2007, a University of Florida team led by Nadya Gorlova used the Spitzer space telescope to study 400 stars that are of similar age to that of the Sun at the time the Moon-forming event took place. They were looking for telltale signs of dust such as would be raised by similar events happening to any planets orbiting these suns. Of the 400 stars studied, only one was accompanied by a dust cloud of the type being sought. From this result, and allowing for the influence of other factors such as the time taken for such clouds to disperse, this team of scientists concluded that at most, 5–10% of solar systems experience the type of collision that formed our Moon.

From this, it would seem that "terrestrial" planets in at least 90–95% of solar systems are more like Venus than Earth. In fact, the percentage may be greater, as not every major collision will be similar to that experienced by Earth. Ours was a glancing blow. A head-on hit might destroy both objects completely. Even exact counterparts of the collision that formed our Moon might not produce similar moons orbiting the alien world. Some of the planets involved might turn out completely different from either Earth or Venus.

Perhaps it is time to speak about "Venusian" instead of "terrestrial" planets in extrasolar systems. It doesn't have the appeal, but it is probably more accurate.

Not all Venusian planets will be exact clones of Venus, of course. We might imagine that, as well as hot Venuses (like our own), there are also cold Venuses orbiting far from the central star (perhaps thrown out into large orbits by migrating gas giants) and temperate Venuses at more modest distances.

"Temperate" may sound appealing, but it does not necessarily imply "benign." These planets would still have a thick crust, incredible episodes of volcanism, and most of the other unpleasant things about Venus; extreme heat excepted. Some of the temperate Venuses may compensate for this deficiency by sporting oceans of sulfuric acid on their surfaces. Not good holiday destinations for space-faring humanity!

Another Black Eye for Jupiter!

Many of us recall those evenings in July 1994 when large fragments of Comet D/1993 F2 (Shoemaker-Levy) crashed and exploded into the atmosphere of Jupiter. After the barrage was over, the planet appeared ringed with a necklace of black spots, which remained surprisingly easy to see even in small backyard telescopes.

At the time, this was presumed to be a rare event and something not likely to occur again within our collective lifetime. Searches of old drawings and images of the giant planet turned up a few earlier candidates for possible giant impact events, but none of these was overwhelmingly convincing, and most astronomers dismissed them simply as Jovian storms of the type well known to all observers of the planet.

Nobody expected history to repeat itself – more or less – during another July just 15 years later!

July 19, 2009, was an unseasonably warm winter's evening as Australian amateur astronomer Anthony Wesley aligned his new 14.5-in. (0.37 m) Newtonian reflector with the brilliant Jupiter, then riding high in the eastern sky. His observing program began at 11 p.m. local time (1300 Universal Time) from his home observatory at Murrumbateman in New South Wales.

After securing several images of the planet, Wesley noted that conditions were starting to deteriorate, and by local midnight he was on the verge of calling it a night. A good thing he changed his mind!

Instead of closing up his observatory, Wesley opted for a 30-min break, after which he planned to assess the sky again and then decide whether to keep observing or call it quits.

Returning to the telescope around 12:40 a.m., he immediately noticed something peculiar about Jupiter's image. There was a dark spot – a *very* dark spot – coming into view close to the planet's south pole. Wesley's first thought was a dark polar storm. However, as the planet's rotation brought the spot into clearer view, the atmospheric conditions at his observing site dramatically improved as if on queue, and the true *blackness* of the spot was better appreciated. No Jovian storm is *that* black!

Wesley then briefly considered the possibility that he was witnessing the shadow of a transiting moon, but this thought was quickly put out of mind. The time and place were wrong, and the spot was far too large. Moreover, it was clear that the feature was moving with the rotation of the planet, rotating in sync with a

Impact scar in the clouds of Jupiter. Courtesy Anthony Wesley.

nearby oval white storm with which he had become well acquainted during earlier observing sessions. He was also very aware that the black spot had not been present when he previously imaged Jupiter just two nights earlier.

The conclusion seemed clear. Sometime between his previous observation on Friday night and the evening of the following Sunday, something pretty large had smacked into Jupiter. This was not your ordinary meteorite. It must have been comparable in size with the Shoemaker-Levy fragments; either another comet or a small wayward asteroid.

Later, it was found that an independent discovery of the black spot was made by T. Mishina in Japan, on an image taken about the same time as Wesley's (i.e. omit "those of" and "images").

Moreover, images obtained at near-infrared wavelength by P. Kalas, M. Fitzgerald, and J. Graham at the W. M. Keck II telescope the following day showed "an anomalous bright [infrared] feature" at the position of the visual black spot. This is in good accord with the interpretation of the spot as being the site of a major impact.

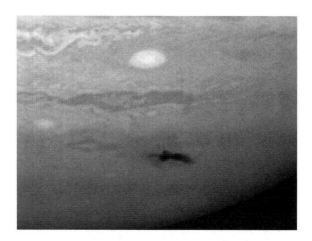

Jovian impact scar 4 days later. Image by NASA/ESA & H. Hammel (Space Science Institute, Boulder Co & Jupiter Impact Team).

News of the event was officially announced to the world by the Central Bureau for Astronomical Telegrams via *Electronic Telegram* No. 1882 on July 22. The spot is there described as having "a complex shape, composed of an impact site with two prominent features separated by about 2° and an ejecta field that extends some 10° toward the west." According to this telegram, the area of the spot is "about 200 million square km."

This event raises an important question. Just how frequently do these impacts *really* occur on Jupiter?

Maybe the occurrence of two such events just 15 years apart is just one of those flukes of statistics that happen from time to time, but it is also possible that their frequency has been severely underestimated in the past.

Only time, and a good watch on Jupiter, will tell.

PROJECT 14
Watching for Jupiter Impacts*

These events are so rare that an observing program specifically watching for them is hardly a good use of observing time. Nevertheless, if you are an experienced Jupiter watcher, there is always a chance that you may be the first person to see the next Jupiter impact, and a brief scrutiny of the planet at the beginning and end of your regular observing program may eventually prove fruitful.

Large impacts such as the one spied by Anthony Wesley quickly draw attention to themselves, but it is possible that smaller ones happen more frequently and may slip by on a casual glance at the planet. Should a quick scan of the planet reveal a small and dense black freckle on the face of the Jovian clouds, waste no time reporting it as a "possible" to the Jupiter Section of your astronomical society, or to a Jupiter enthusiast known to you. Small events are probably short lived and, if a discovery is to be verified, time is of the essence.

4. Weird Meteors

Our Solar System may appear to us as a vast emptiness. Yet, the spaces between the major planets are really far from empty. Besides the large numbers of so-called "minor bodies" (asteroids and comets of various types), millions upon millions of even smaller objects, ranging in size from dust motes to boulders, drift through the cosmic wastes. These are actually fragments of the minor – and even of some of the major – citizens of the Sun's family. For the most part they remain invisible to Earth's inhabitants until one happens to strike our atmosphere. Then, for a few moments, that object may become the most conspicuous object in the sky. Sometimes, just occasionally, one such object may be reported as doing something really weird!

Fiery Darts in the Sky

Anyone watching a relatively dark and unobstructed sky for 10 min or thereabouts will almost certainly see at least one star-like point of light dash across the heavens for several degrees before vanishing into oblivion. One may naively imagine, as sometimes young children do, that one of the mighty stars has lost its place in the firmament and fallen to destruction. Indeed, the popular name "falling star" (and to a lesser degree its alternative "shooting star") appears to reinforce this idea.

Nevertheless, the truth is far less apocalyptic.

Although these dashing points of light may look like stars, they could hardly be less similar to the true luminaries of the cosmos. Rather than great balls of glowing gas at immense distances from Earth, these "falling stars" are nothing more than tiny specs

D.A.J. Seargent, *Weird Astronomy*, Astronomers' Universe,
DOI 10.1007/978-1-4419-6424-3_4, © Springer Science+Business Media, LLC 2011

of natural space debris burning up in a brief flash of incandescence in our atmosphere, a mere 100 miles or thereabouts above our heads. A fleck of rock the size of a rice grain, or even a speck of coarse sand, is enough to cause a typical "falling star." Something the size of a tennis ball entering our atmosphere blazes into a brilliant fireball.

The realization of what these strange moving lights really are did not come quickly. For a long while, they were thought to be exclusively atmospheric phenomena, most probably related to lightning. This belief is not difficult to understand. They are brief and very often dart across the sky at high speed. Very large ones are even followed by peels of "thunder." (Actually, sonic booms generated as these larger-than-normal bodies penetrate into the denser, lower, levels of the atmosphere.) Even the technical name for these objects – "meteors" – still reflects this belief in an atmospheric origin. Note that "meteorology" is the name given to the study, not of meteors but of Earth's atmosphere and weather.

A few dissenting voices were raised however. For instance, Edmond Halley (1656–1742), an astronomer not known for his ready acceptance of orthodox opinions, held that these objects were of extraterrestrial origin, but his opinion was ahead of its time and won few converts. The real turning point in people's opinions of the true nature of meteors was largely precipitated by an unusual and remarkably spectacular occurrence, about which a little background should be given.

Meteors, as already mentioned, can be seen on any clear night if one watches the sky for long enough. Radar echoes show that they are just as frequent during the daylight hours, so the number striking our planet on any day is really very large. However, there are times when this number is significantly enhanced – when meteors appear in the skies at rates far higher than average. These events are termed "meteor showers" and take place when Earth passes through a stream of particles moving around the Sun along similar orbits. By the way, these particles are known as "meteoroids" while they are traveling in space and not incinerating themselves within the atmosphere of Earth or some other planet.

Many "showers" of meteors occur during the course of a single year. Most of these are regular annual occurrences, so meteor watchers are ready for them. However, it is also true to say that the word "shower" is less than descriptive of many of these events. If that word conjures up images of meteors streaming from the sky in a constant barrage, then forget it! Events of this type do occur at rare intervals but are more often described not as "showers" but as "storms." But more of this in a little while.

Nevertheless, the strongest of the annual "showers" do manage to drop meteors at respectable rates, though in no way comparable with the very rare storms. An individual observer may see between 60 and about 100 meteors per hour in showers such as the Perseids in August and the Geminids in December. These may indeed live up to the title of "showers," but some other annual events are not so impressive. For some of these minor streams, a single observer would be lucky to count one or two "shower" meteors in an hour!

A significant feature of "shower" meteors is the fact that they appear to radiate from a very small region of the sky. This is termed the "radiant," and each meteor shower is named for the constellation in which its specific radiant lies. Thus the two showers mentioned above – Perseids and Geminids – have their radiants located in the constellations of Perseus and Gemini. The first of these, by the way, also carries the nickname of "the Tears of St. Lawrence" due to its occurrence around the time of the feast day of that saint.

The radiant is, however, merely an effect of perspective. In reality, it signifies that the meteors arrive in the vicinity of Earth along parallel – or nearly parallel – orbits. A similar effect is responsible for the apparent convergence of railroad tracks. Also, when driving through falling rain or snow, it is noticed (preferably by the passenger and not the driver!) that if one looks ahead of the vehicle and upwards, the raindrops or snowflakes appear to diverge from a single point. Again, this is due to the relatively parallel paths of the drops or flakes as they fall through the atmosphere and encounter the forward moving vehicle.

On rare occasions, an intense meteor display will drop hundreds of meteors in a single hour. In the most extreme instances, where meteor rates exceed 1,000 per hour, the term "meteor storm" is used to distinguish the event from "normal" showers. These storm events are very rare, but they afford some of the most spectacular and even terrifying sights seen in the night sky. Note, for instance, this reaction to one such occurrence observed in Portugal on October 23 (the 30th on the Gregorian Calendar), 1366:

> There was in the heavens a movement of the stars such as men never before saw or heard of ... they fell from the sky in such numbers and so thickly together that ... the air seemed to be in flames, and even the earth appeared as if ready to take fire.

Just such a meteor storm was seen on November 11, 1799, by the great Alexander von Humboldt during his exploration of South America. Humboldt recorded that on that night he witnessed "thousands of meteors and fireballs moving regularly from north to south with no part of the sky so large as twice the Moon's diameter not filled each instant by meteors." From this description, Gerald S. Hawkins calculated that if the average meteor path was 10° long, from Humboldt's description, an individual observer would see, in any one instant, some 200 meteors (assuming an observers' field of view of around 2,000 square degrees). This translates to an hourly rate of 720,000!

Realistically, the assumption of 10° for the average path of these meteors seems a little short. Probably something in the order of 30° is more realistic, which reduces the total rate to about 240,000 per hour – by all estimates, still an incredible display. Humboldt also seems to have acquired some information from the inhabitants of the area about another spectacular storm of meteors that had occurred over the same region some 30 years earlier.

Then, 34 years later, there occurred one of the most important events in the history of meteor astronomy. On the morning of November 12, 1833, meteors fell "as thick as snowflakes" over the central United States. People awoke in the early hours of morning finding their bedrooms illuminated by continuous flashes. One story tells of a child bursting into his father's bedroom telling him that the sky was on fire!

By all accounts, the storm dropped at least 100,000 meteors per hour as seen by an individual observer.

It was during this storm that the phenomenon of a radiant was observed for the first time, and its significance quickly realized. Whatever was causing these fiery streaks in the sky was arriving along parallel paths. This was no atmospheric phenomenon akin to lightning but something genuinely astronomical

– although not everyone was quickly convinced. (At least one scientist with a passion for meteorology strongly resisted the attempt by astronomers to "steal" one of his subjects!)

A search of historical records confirmed the occurrence of great meteor storms every 33 years. The most straightforward explanation was the existence of a swarm of particles traveling around the Sun along a comet-like orbit with a period of about 33 years. This was the same swarm that encountered Earth in 1799 as witnessed by Humboldt. We now know that it was also the one responsible for the great storm of 1366, a brief description of which was quoted earlier. Having a radiant in the constellation of Leo, it is now known as the Leonid meteor shower and has been responsible for the greatest meteor storms of recorded history.

Following the 1833 storm, meteor science matured rapidly. Annual showers such as the Perseids and Lyrids were noted and a prediction made for the return of the Leonid storm in the late 1860s. This prediction was fulfilled, although the storm was not as great as that of 1833. Furthermore, the Leonids were also recognized as an annual shower, with a weak return each November. The display during the "off" years is, however, very different

Leonid fireball, November 17, 2002. NASA-ARC/G. Varros.

from the 33-year blizzard. Hourly rates sink to ten or less well away from the years of the intense returns.

By that time, however, a comet (Tempel–Tuttle) had been discovered sharing the same orbit as the Leonids, and others were found moving along the paths of the Perseids and Lyrids. The link between comets and at least some meteors had been found!

Another Leonid storm was forecast for 1899, but, alas, this prediction was not fulfilled. A nice shower did occur eventually, but rates did not approach storm intensity, let alone the incredible deluges of 1799 and 1833. Likewise, 1933 failed to produce a true storm, though hourly meteor rates in the low hundreds were reported from some parts of the world, and one observer likened the meteors to sparks from a "sparkler," or hand-held firework.

Consequently, as 1966 approached, hopes of seeing anything too spectacular were less than high. Annual rates had fallen to as low as four or thereabouts by the late 1950s, and most astronomers thought that the main swarm had been deflected forever away from the earlier Earth-crossing orbit. Moreover, Tempel–Tuttle had not been observed during its returns of the late 1890s and early 1930s, and there was some thought that it may have faded out and ended its meteoroid-producing career.

It came as a surprise, therefore, when the Leonids made a surprisingly snappy recovery in 1961, although the rates were nowhere near storm intensity. The following years were not so lively, but another surprise followed with the recovery of Tempel–Tuttle itself in 1965. That year, the Leonids also put on a truly spectacular show, raising some cautious hopes that 1966 might not be a complete fizzler after all.

That turned out to be an understatement. The mighty meteor storm that burst forth in the central United States that year has been reckoned by some as the greatest ever seen. During the peak 20 min of activity, meteors fell at the rate of some 40 per *second* as estimated by an individual observer!

The most recent Leonid storms occurred in 1999, 2001, and 2002. Weak compared with the 1966 event, they still dropped several thousand meteors per hour, but they differed from former events in having been predicted with a precision undreamt of in earlier years. In the late 1990s, astronomers Rob McNaught and David Asher developed a rather precise model of how meteoroid streams, deriving from comets, evolved, and by using this model they were able predict with remarkable accuracy just when, where, or if denser filaments within the stream would encounter Earth. These denser filaments are the ones capable of producing unusually rich meteor showers or meteor storms.

McNaught and Asher predicted the time, place, and approximate intensity of the 1999 storm (allowing astronomers to travel to the predicted target area, the Middle East, and observe it), the lack of a storm in 2000, and the return to storm levels in 2001 and 2002. All of these predictions were fulfilled.

If there is a downside to the Asher–McNaught discovery, it is the date of the next predicted Leonid storm – 2098 – by which time most of us will not be worrying about meteors.

Other meteor storms beside the Leonids have occurred, but they are very rare phenomena, and, alas, there is no indication of any more in the foreseeable future. During the last two 200 years, the only meteor storms other than the Leonids have been the Andromedids (associated with the now defunct Comet Biela) in 1872, 1885, and (marginal storm rates) 1892, the Draconids (associated with Comet Giacobini-Zinner) in 1933 and 1946, and very brief storm levels during the strong return of the November Monocerotids in 1925 and again 10 years later. With the possible exception of the latter (which appears to be a somewhat atypical stream that also gave quite a strong return in1995), it is not likely that there will be any unusual activity from these in the years ahead.

There is much more that could be said about meteors and meteor showers, but this would take us beyond the scope of this book. After all, meteors and meteor showers are no longer "weird," even if they may have been thought such just a couple of centuries ago.

Nevertheless, we can see from the above brief historical account just how recent the serious study of meteors is, and it is not surprising to find that there is still much to be known, much to be explained – much, we might say, that remains "weird." In the pages ahead, we will look at some of the oddities reported by meteor observers. Incidentally, the oddities looked at here are those of meteors, not meteorites. We look at a few remarkable cases of the latter in Chap. 7. The great majority of meteors originate from comets, as

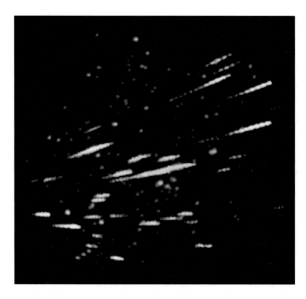

Meteors stream away from the constellation of Monoceros during a brief but intense outburst in 1995. S. Molau and P. Jenniskens. NASA.

we have said. By contrast most, if not all, meteorites appear to be fragments of asteroids or even of larger bodies such as the Moon and Mars, and, although their flight through the atmosphere is accompanied by "meteoric" phenomena, the scale of this is of an altogether different order from that of the typical shooting star.

Double or Parallel Meteors

These are meteors that appear together or in very quick succession, with the second following the first along the same trajectory like a child following its parent. During very rich showers or storms, this may not seem especially weird (it was noted during the Leonid storm of 2001, for instance). But when there is no storm, the chance of two random meteors coming in through the atmosphere like this is, well, astronomical! Yet it does happen. Not very frequently it is true, but often enough to stretch the probability of a mere chance occurrence.

So what is happening here?

The most credible explanation is that the two meteors were until very recently joined together as a single body that split apart just prior to entering Earth's atmosphere. As to why such a split should occur, it is known that comet nuclei not infrequently split for no apparent reason, so maybe their smaller siblings suffer the same, apparently random, fate at times.

Just a thought, probably a crazy one, but it is known that passage through Earth's auroral zones can induce static electric charges in certain large and loosely compacted meteoroids, and that this sometimes causes them to burst apart into clouds of dust. These clouds have been observed as dust showers by artificial Earth satellites launched to monitor cosmic dust particles in Earth's near environment. From these observations, the masses of the parent meteoroids have been estimated and found to be very close to those of typical bright fireballs. Maybe similarly induced charges can at times split some meteoroids without causing total disruption.

While talking of parallel meteors, occasionally two shooting stars having very similar characteristics will be seen moving along apparently parallel trajectories, either at the same time or in quick succession. This author's own most striking experience of this was of two faint meteors of short and truly parallel trajectories, looking almost like clones of one another. It was hard to imagine that they were not in some way related, although the parallel and short nature of their paths suggested two separate radiants just a few degrees apart. By contrast, if they had entered the atmosphere on *truly* parallel orbits, perspective would have caused them to apparently diverge from a single radiant point, as explained earlier. Maybe instances such as this are purely coincidental ... or could they be the result of a meteoroid splitting well in advance of entering Earth's atmosphere?

Erratic Meteors

"Erratic" is a blanket term covering a multitude of oddities. And some of the behavior reported for meteors certainly fits that bill.

At the more readily explicable end of the spectrum, we have events such as the ziz-zag meteor observed by Perry Vlahos of the Victorian Astronomical Society in Australia. This one appeared to execute a sharp change of direction mid flight, hence the description of a "zig-zag path."

Apparent explanation – the meteor split during flight, with the fragment hiving off at a sharp angle to the original trajectory.

Following this disruption, the original meteor quickly faded out, but the fragment continue along its (new) path, giving the appearance of a single object executing a sharp turn.

Actually, meteors making several zig-zags have been explained in this way, but we may raise the question as to how many zigs and zags a meteor must perform before the explanation of multiple fragmentations, fading primary fragments, and surviving secondary ones starts to become a little stretched!

Vlahos also recalled witnessing another erratic meteor, the explanation for which is not so forthcoming.

This one started out conventionally enough, heading straight downward like your typical falling star. Then, before fading, it appeared to bounce back upward like a yo-yo coming to the end of its string!

That is surely weird. As we will see in a later chapter, the so-called autokinetic effect can sometimes cause stars seen close to the horizon to appear to bounce up and down, but this does not seem to be the situation here. Maybe the splitting meteor explanation can again be called into play, except that this time the fragment appeared to shoot upward from the primary just as the latter faded away. This upward motion need only be apparent. It might simply be the result of perspective, but the direction only needs to *appear* real to give rise to the bouncing meteor effect.

By the way, speaking of meteors moving upward, if you ever witness a very large fireball – especially one accompanied by hissing sounds (more on this below) and followed by loud sonic booms – apparently heading upward in the sky, get under the nearest table! This is probably a meteorite, and its apparently upward motion, due entirely to perspective, means that you are in the fall zone!

There is another odd type of meteor trajectory that has been reported from time to time and that could justifiably be termed erratic. This is the curved path. Curved paths are reported frequently enough by experienced meteor observers to have aroused controversy in meteor observing circles. They are too common to be dismissed out of hand but too odd to fit neatly into what "should" happen.

According to some meteor experts, the reported curved paths are simply the results of optical illusion. In support of this explanation, it is noted that curved meteors are more frequently seen in an observer's peripheral vision. The favored explanation is that the meteor catches the observers' attention by entering the periphery of his field of view, but, once alerted to its presence, the observer quickly turns for a better view, and it is this quick turning of the head that briefly disorients him/her and makes the meteor's path appear curved.

This seems a reasonable and simple explanation, but there is a catch. Not every curved meteor is seen at the periphery of vision. In fact, some experienced observers have noted curved trajectories for meteors appearing almost centrally to their field of view. For these instances, at least, the above explanation starts to look a trifle tenuous.

On the other hand, not everything is as it seems. Meteor observer Pierre Martin recalls plotting a very peculiar object during the annual August Perseid shower. Apparently originating from the shower radiant, this one differed from its fellows by following an amazingly curved trajectory – and not simply curved but S-shaped! Now that was truly an amazing meteor.

But *was* it really a meteor?

Actually, no. It was simply a bird reflecting the glare of streetlights! Its true nature became apparent a few nights later when Martin noted a similar object performing the same type of maneuver. This time however, its true identity was readily unmasked, much to the observer's disappointment. In fact, since that incident, Martin seems to have rejected the possibility of curved meteors altogether, assuming that those reports not arising from the above-mentioned illusion are simply nearby objects such as birds or bugs illuminated by ground lighting.

Although there is no doubt that such things are mistaken for curving meteors from time to time, as his own experience proves, it is difficult to agree that all sightings can be dismissed so easily. What, for example, can be made of the following report found in the *Boston Globe* of February 23, 1990?

> Reports of a fireball that blazed through the skies over the Northwest on Sunday, changing colors and <u>even executing a fiery loop</u> before vanishing, have been filtering into local agencies…[Underlining mine].

Observers from Nova Scotia to New Jersey reported the spectacular fireball, which they said was visible for more than 10 s at 7.30 p.m. Sunday in the southeastern sky.

This was clearly not a bird, even though its reported flight path was highly anomalous for a meteor!

While not exactly "curved," some meteors have been said to follow a more subtly "wavering" trajectory. Although the shifts from straight line motion are only small, it is difficult to understand how an object plunging through the atmosphere at the speed of a meteoroid could endure even the smallest alteration in direction. Then again, any change in direction large enough to be seen by the naked eye from ground level is *not* altogether miniscule.

It is tempting to explain this phenomenon in terms of optical illusion or atmospheric refraction, more or less analogous to the twinkling of stars. Although one or both of these suggestions might explain some observations, they do not appear adequate to account for them all. This phenomenon was even photographed in August 1997 when a "wavering" Perseid was caught tracking through Capricornus. What *is* really happening here?

It has been suggested that the meteoroids might be rapidly spinning in these instances. If such were the case, this may also explain the appearance of corkscrew-like tails associated with some meteors.

An early record of one of these appeared in the journal of one Thomas Hughes, an officer serving under General Burgonyne in the American Revolutionary War. On the evening of November 21, 1779, Hughes noted that "A strange meteor was seen in the south, just as the Sun went down. It appear'd (sic) like a ball of fire and left a long trail of light – something like the turnings of a corkscrew – visible for near an hour." Its long period of visibility undoubtedly meant that the trail was composed of dust and, being so soon after sunset, remained lit by the Sun's rays as it very slowly dispersed in the upper atmosphere.

Sometimes as a meteor recedes from the observer, a combination of increasing distance and end-on perspective may make it appear to slow down. An extreme and atypical instance of this was described by Martin for a bright meteor, apparently a member of the Geminid shower, on December 13, 1988. As described, this object moved quite rapidly away from the direction of Gemini and beyond Orion, descending toward the horizon. However, as it descended, it seemed to both slow and brighten until it appeared almost to stop and take on the appearance of a yellow-white ball of light equal in luminosity to the planet Venus. Then, it simply blinked out and was gone.

The weirdness of this observation concerned not just the apparent slowing of the meteor but its increase in apparent size and brightness with increasing distance from the observer. It must have provided quite a show for anyone fortunate enough to have been near the end point of its trajectory.

Black Meteors

Have you ever been watching the sky for meteors – probably during the time of a meteor shower – when you suddenly seem to see a dark object shoot across the sky for a short distance. What you see looks just like a small meteor, in all but one important respect. It is not luminous!

"Black meteor" reports are certainly not confined to novice meteor watchers. They are experienced by old hands at the art as well. The trouble is, according to all accepted wisdom, they simply cannot exist!

Any macroscopic physical object traveling at meteor speed through the atmosphere necessarily *must* become incandescent due to friction. There is simply no way around this. So if they don't exist, why are they so often reported?

The usual explanation is that these events are optical illusions. In support of this, it is noted that many of the sightings occur either just after a meteor watch begins or shortly before one ends. In the first instance, it is asserted, the human eye has not properly settled down to dark adaptation, and misperceptions are more likely to happen.

Conversely, toward the end of the watch (especially if it has been a long one), fatigue is prone to start setting in. Under these circumstances, one is more likely to "see" something that isn't really there.

On the other hand, sometimes a black meteor might indeed be a real object. The trouble is, it will not be a meteor!

If one is observing from a site that is free from low-level glare while at the same time having a mildly light polluted sky – a rural area close to a large town, for instance – a flying insect silhouetted against the background sky might be mistaken for a dark meteor. In this instance, however, the "track" is likely to be quite long.

If there is also a degree of local glare, as is common on a lighted suburban street, a flying insect has less chance of being seen in silhouette against the sky. Under these conditions, it is more likely to be illuminated from beneath by the local streetlights. Rather than being mistaken for a black meteor, it is then far more likely to be mistaken for an ordinary luminous meteor, as is evidenced by the earlier instance of a night bird being mistaken for an erratic Perseid!

PROJECT 15
Black and Erratic Meteors

Here is an interesting project for somebody with meteor-watching friends in both suburbia and the country. Simply have them record each instance of apparent black or curved meteors and compare the two sets of results. If bugs and birds are really responsible for these reports, the rural observers should report more black and fewer curved "meteors" than their suburban counterparts.

Does this happen? Only one way to find out!

Nebulous Meteors

The typical appearance of a meteor is that of a small point of light or, in the case of a fireball, a bright and often teardrop-shaped object of more or less high intensity. But on rare occasions, a meteor observer will see something quite different, something looking more like a fuzzy comet shooting across the sky!

Variously termed "fuzzy meteors," "diffuse head meteors" or (more frequently) "nebulous meteors" these strange objects are generally rather faint (in the second or third magnitude range) and show as a fuzzy ball about half the size of the full Moon. Some are even larger and fainter. One reported from western Australia several years ago was described as a faint nebulous patch of light, about as large as the full Moon and shining with a total light equivalent to a fourth magnitude star, albeit moving with a speed and duration similar to that of an ordinary meteor. Presumably these strange meteors are caused by extremely friable objects that dissolve into clouds of smoke-like particles upon entering Earth's atmosphere.

Even telescopic nebulous meteors have been reported. For instance, well known nineteenth century astronomer F. Winnecke recorded the following notes:

> *November 23, 1881. A grayish elliptic cloudlet 2' in diameter, tolerably well defined, shot exactly across k Aquarii. Its brightness equaled that of Herschel's first class of nebulae* [i.e., somewhere between magnitudes 9 and 12, though probably closer to the brighter value], *but it looked more compact and less transparent.* [Winnecke also noted that on August 31, 1880, "A very diffused nebulosity of this class" and having a diameter of "many minutes" was seen in Delphinus and appeared "decidedly reddish" in color.]
>
>   Sept 14, 1882. A faint grey shadow passed over S Persei.
>
>   July 28, 1883. A faint grey perfectly diffused vaporous mass, 4' in diameter ... moved in a curvilinear path out of the [telescopic] field. [A curved telescopic nebulous meteor!!]

A curious feature sometimes reported in relation to the more familiar naked-eye nebulous meteors is their apparent "nearness" to the observer. They do not have the appearance of relatively distant bodies passing through the upper atmosphere, but instead give the impression of being "foreground" objects quite close to the observer.

This is almost certainly an optical illusion. At night, depth perception can easily become muddled, and it is possible that the diffuse appearance of these meteors suggests something "out of focus" and thereby very close to the eye of the beholder. This suggestion is purely a guess, but it is something that should be capable of being followed up experimentally. Indeed, there may be some evidence that large diffuse objects seen against the night sky can give the illusion of being in the lower atmosphere. In 1996, the very bright comet Hyakutake passed relatively close to Earth, and for a while appeared high in the night sky as a very large diffuse mass sporting a glorious tail. At least one observer commented that the comet gave the appearance of being somehow close at hand rather than at truly astronomical distances. Maybe this illusion is exacerbated in the case of nebulous meteors by the latter's swift motion.

Another peculiar feature of these meteors is the apparent dearth of truly bright ones. We never seem to hear of a nebulous fireball ... or do we?

Western Australian meteor observer Jeff Wood wonders if such events *do* actually occur, but if they simply fail to be recognized for what they are. Indeed, they may fail to be recognized as meteors at all!

Wood asks what a nebulous fireball would really look like. Imagine a very bright but cloudy mass traveling at the speed of a meteor – probably a slow meteor – and seen against a dark sky. Most likely, such an apparition would take the form of a cigar-shaped shimmering mass of light, probably sporting a luminous trail for some distance across the sky. It would not look like a normal fireball, yet it would be conspicuous enough to attract the attention of folk who would not normally notice meteors, still less, nebulous ones.

What would these folk report?

A UFO perhaps?

Wood thinks so and suggests that meteor observers should pay attention to those UFO reports involving cigar-shaped objects cruising across the night sky. Maybe, hidden among sightings of the planet Venus, bright 'ordinary' meteors, ball lightning, and a host of other luminous phenomena are instances of bona fide nebulous fireballs. If he is correct, a very interesting phenomenon may be lurking in lists of observations seldom scrutinized by meteor experts.

Crackling, Popping, and Hissing Meteors

Most of the meteors we see streaking through the night sky complete their careers in silence. Larger objects, especially those massive enough to reach the ground as meteorites, are a different story. Very large bodies, such as the (fortunately!) very rare crater-blasting meteorites penetrate Earth's atmosphere with such ease that they strike the ground with cosmic velocity and end their days in a violent explosion. The more common types of meteorites, however, are sufficiently checked by atmospheric resistance while still several miles above Earth's surface and make their final descent at the speed of any object falling from a great height. The fireball accompanying them fades out, and the final segment of their journey is made as dark bodies, rapidly cooling as they approach the ground. Stone meteorites are generally slightly warm if picked up immediately after their fall, though iron ones may briefly be a little too hot to handle. None will be glowing, and there have even been instances of frost forming on freshly fallen stones!

Nevertheless, the falling meteorite retains supersonic velocity far enough into the lower atmosphere to give birth to a cannonade of sonic booms. But because sound and light travel at vastly different velocities, these are not heard at the same time as the fireball is seen. It is like thunder and lightning. Despite the instantaneous lightning and thunder of Hollywood storms, we all know that if we hear the thunder at the same time as we see the lightning, it is too close for comfort. The more distant the lightning strike, the longer the interval between seeing the flash and hearing the thunder.

Very bright meteors are sometimes visible in broad daylight, like this fireball seen over Spain on January 4, 2004. Courtesy Salvador Diez and Spanish Fireball and Meteorite Recovery.

The same is true of a meteorite. The normal sequence of events is to see the fireball and then, after this has faded out, to hear the sonic booms. The difference in time may be considerable if the meteorite is quite distant.

There is nothing weird in this. The sounds accompanying the nearby fall of a meteorite may be frighteningly loud (a bright fireball that flew over Shepperton in Victoria was reported to have burst water pipes!), but they are exactly as expected for an object entering the lower regions of the atmosphere at supersonic velocities.

What *is* weird – or was considered such for many years – are the recurring reports of *simultaneous* sounds accompanying some bright meteors. These alleged sounds are quite different from the thunderous sonic booms we have been discussing. They are more in the nature of crackles, pops, rattles, and hisses – often compared to the sound of water being dropped onto hot iron.

Sometimes, such "anomalous" sounds have been reported for meteors that did not produce the more familiar delayed thunderous noises. Thus the English astronomer W. F. Denning reported that a witness to one of the three bright meteors (probably members of the Alpha Scorpiid stream) seen on June 7, 1878, "fancied he heard a crackling sound" accompanying the meteor's flight. No other sound is mentioned, and we may wonder if there is a hint of skepticism in Denning's use of the word "fancied."

These strange sounds have been recorded for a long time. Russian meteorite expert E. L. Krinov finds this account in the *Lavrentenko Annals* of 109 AD:

> A large serpent falling from the clouds ... all this time the Earth was rattling.

Krinov suspects that this account refers to a meteorite fall accompanied by "rattling sounds," although the statement is a little too vague for us to be certain of this. The mention of a "serpent falling from the clouds" may even be an account of a descending tornado funnel. We simply do not know.

We are on far safer grounds with the anomalous sounds reported to accompany a bright fireball seen in England in 1719. Edmond Halley, on hearing about these events, correctly realized that they could not be sounds generated by the meteor itself, as it was simply not possible for sound to be transmitted instantaneously over the distances required. If this sound was generated by the meteor, it should arrive at the same time as the sonic boom.

What was reported *could* not happen, and therefore, according to Halley, *did* not happen. He fell back on that old chestnut explanation for things that "can't" happen according to the current scientific knowledge – psychology. The effect was psychological only.

Another early instance occurred in 1784 when a large fireball was observed from England and Scotland all the way to continental Europe. Once again, this was reportedly "heard" by many observers. Similar to the earlier instance, this "sound" was also described in terms of a "hissing" and, as before, occurred simultaneously with the sighting of the fireball.

Commenting on this latter instance, Royal Society Secretary Dr. Thomas Blagdon, like Halley before him, leaned toward the idea that the sounds were the product of "an affrighted imagination," although, unlike Halley, he left the door open to alternative explanations. Impressed by the veracity of many of the witnesses, he refrained from dogmatism and suggested that the anomalous sound was something that may only "be cleared up by future observers."

Despite this small concession, the psychological explanation became scientific orthodoxy for the next two centuries. That is not to say that the occasional scientific "heretic" didn't raise a voice against it, but their voices were quickly drowned out by majority opinion.

One such dissenting voice was that of Professor J. A. Udden, whose investigation of the great Texas fireball of 1917, and the anomalous sounds which were reportedly associated with it, led him to wonder if the true explanation for these sounds might not be "sought in ether waves that, on meeting the earth, or objects attached to the earth, such as plants or artificial structures, are in part dissipated by being transformed into waves of sound in the air."

A very similar explanation was proposed (apparently independently) by Elmer R. Weaver to meteorite expert H. H. Nininger in 1934. Nininger was apparently sympathetic to this approach and is quoted as saying that he thought the solution to the phenomenon was "a problem of physics rather than psychology."

Another scientist who took the "physics" rather than the "psychology" approach was Professor Peter Dravert of the Omsk

University. It was Dravert who coined the term *electrophonics* to describe the phenomenon, although the word did not become a common part of the meteor observer's lexicon until far later. Dravert's compatriot, the great meteor expert I. S. Astapovich, also took the "physical" approach and actually spent much time studying the phenomenon.

Nevertheless, it was psychology that continued to hold sway. Superficially, the psychological hypothesis appeared to have several points in its favor.

Probably the most important was the fact that, as already mentioned, the anomalous hissing sound arrived at the same time as the light from the meteor. As sound can never speed up to that of light, nor light slow to the speed of sound, this appeared to make the phenomenon physically impossible.

Secondly, interviews with witnesses revealed that not everyone in a group heard the sounds. Sometimes several people would hear it, but at other occasions the same group would be divided, some hearing the hissing sound and the others denying that anything of the sort had occurred. This was noted among witnesses of, for instance, the Karoonda Meteorite, which fell in southern Australia on November 25, 1930.

Thirdly, the psychological explanation seemed to have an explanation as to why this supposedly subjective sound took the form of a hiss or crackle. It was like a firework! According to this suggestion, because a bright meteor looks somewhat like a firework, the mind is tricked into filling in the missing sound effects.

It would have been interesting to see if people not acquainted with fireworks (very young children perhaps, tribal natives, or folk living in very remote regions) ever experienced these anomalous sounds.

Be that as it may, there were more immediate problems. If the sounds truly were subjective reactions to something that looked like a firework, they must always follow the very first glimpse of the meteor. However, there were credible reports of the witness's attention having been drawn to the fireball *after first hearing* the sound.

For example, one witness of the great Murchison Meteorite fireball of 1969 mentioned that she was in the garden when her attention was drawn by a loud hissing sound, which she described as sounding like the tires of a vehicle being driven over a wet road. Looking up, she *then* saw what she at first thought was the Sun, though immediately realizing that she was not looking in the Sun's direction. The fireball burst and vanished, and short time later there came a sound like thunder, only far louder. Despite making it plain that her attention was first aroused by the hissing sound, this witness was nevertheless informed that her experience was purely subjective and psychological. For that to have any ring of truth whatsoever, her sense of remembered time would need to be reversed, for which there was absolutely no evidence. In order to fit the psychological model, this witness's report needed to be twisted into more contortions that a fireball's dust trail!

The real breakthrough came in 1978, when a fireball some 40 times brighter than the full Moon shot across the skies of eastern Australia, crossing the east coast between the cities of Sydney and Newcastle about 90 min before sunrise. (This, incidentally, is where the present writer lives, and the meteor must have passed almost directly over my house. Actually, I had been observing earlier that morning but went back to bed just before the meteor arrived. Such is life!)

This meteor was widely seen from Sydney to areas north of Newcastle, and the event was thoroughly investigated by Colin S. L. Keay, Physics professor at the University of Newcastle. Initially it was hoped that an investigation might lead to the recovery of meteorite fragments, but it quickly became apparent that anything that might have reached ground level would have plunged into the Tasman Sea and finished up fathoms down in the ocean. The ultimate importance of Keay's investigation, though, came from an entirely unexpected direction.

In report after report, as collected by Keay, mention was made of sounds. Not the sonic booms arriving after delays of minutes, but a variety of noises heard simultaneously with the sighting of the fireball.

Keay's investigation kept turning up statements such as:

"… a noise like an express train or a bus traveling at high speed. Next an electrical crackling sound, then our backyard was as light as day"

"A noise could be heard. A low moaning, swooshing."

> "I heard a sound like an approaching vehicle and saw a flash
> of light as everything was lit up like daylight."
> "It was a loud swishing noise."
> "heard a noise like a 'phut.'" (a second person standing next
> to this witness heard nothing)
> "a sound like steam hissing out of a railway engine."

These reports came from witnesses who were spread out over 100 miles along the East Coast. They had certainly not collaborated with one another prior to being interviewed.

Keay's lingering suspicion that something more interesting than psychology was involved here now seemed entirely justified, and he set about to afford the phenomenon the in-depth investigation that was so long overdue. Even then, most meteor experts to whom he spoke told him that he was wasting his time studying this subject!

Keay opined that, if the sounds were real, there was only one way the energy causing them could be transmitted so quickly over such large distances. The culprit had to be electromagnetic radiation of some type or other.

Yet, prior observation of bright meteors had shown that these events do not generate radio waves, and there seemed to be no way that the radiation which they clearly do produce – light and heat – could be perceived as sound. (Actually, there *is* one exception, namely, a rare psychological/physiological phenomenon called synesthesia, where there is a fusion of perceptions, and people can literally "hear sights and see sounds." There is no reason, however, to think that only synesthetics experience anomalous meteor sounds.)

Other than the visual region, the only part of the electromagnetic spectrum where meteors *may* emit radiation is in the band from about 1 Hz to 100 kHz. There is no direct evidence for radiation at these frequencies, but neither is there evidence that it does not – or could not – occur. The fact that this range of frequencies covered audio frequency was, at least superficially, very interesting. The task that Keay set himself essentially came down to searching out a mechanism by which meteors could generate electromagnetic radiation within this band of frequencies.

He first considered the intense burst of emission produced by nuclear explosions. As any fan of the James Bond movie *Goldeneye* knows, nuclear blasts can perturb the geomagnetic field to such a degree that a burst of electromagnetic radiation capable of burning out electronic equipment is generated. Reportedly, this can be heard as a "click" by witnesses in bunkers relatively close to such a blast. A large meteor exploding might generate a similar effect.

Keay also noted that a paper published in 1965 by V. V. Ivanov and Yu. A. Medvedev showed that a large meteor entering the atmosphere at a fairly steep angle could disturb the normal geo-electric field, resulting in electrostatic discharges whose effects might be audible as brief "swishes." This seemed a possible candidate for swishing sounds associated with meteors seen passing directly overhead.

Nevertheless, the big problem lay in explaining how a meteor could generate sustained electromagnetic radiation. No obvious process was known, which, of course, is why the whole thing had been swept under the psychological carpet since Halley's day!

In a flash of inspiration, Keay hit upon a theory of sunspots developed by English astrophysicist Sir Fred Hoyle. Briefly, Hoyle theorized that sunspots occur when energy is trapped in twisted magnetic fields. What might happen, Keay pondered, if Earth's magnetic field similarly becomes trapped in the turbulent trail of plasma left by a bright meteor and is released when the plasma cools and the ionization neutralizes itself?

His calculations confirmed his suspicion. A "magnetic spaghetti" could indeed arise within turbulent meteor trails, but only when the meteor and its trail are relatively low in the atmosphere. For the effect to be sustained for 10 s or longer, the meteor must also arrive along a shallow atmospheric trajectory. This fits well with the fact that (contra Ivanov and Medvedev) meteors in shallow, low trajectories seem to be the ones most frequently associated with reports of anomalous sounds.

Keay's work was published in the journal *Science* in 1980, and in 1983 the concept was expanded by V. A. Bronshten, who demonstrated that a fireball twice as bright as the full Moon could generate well over a megawatt of radio power by the "magnetic spaghetti" effect. Subsequently, T. Okada and colleagues in Japan were able to directly detect radio waves of the required frequency from a bright meteor.

Keay's next step was to reproduce the effect under laboratory conditions. His initial experiments, performed during a visit to the Physics Department of the University of Western Ontario, revealed some very interesting factors. By subjecting a number of volunteers (including himself) to electric field variations, he found that three of his subjects were unusually sensitive to these variations. The reason seemed to be their hairstyles! Two of the volunteers were female and wore Afro-style haircuts. The third was a male with very long and soft hair. Apparently, their hair acted as antennae!

Keay experienced this effect first hand. He found that his own sensitivity increased when he kept his glasses on during the experiment!

Continuing the experiments when back in Newcastle, with the assistance of graduate physics student Trish Ostwald, Keay found that one's sensitivity – one's ability to hear the anomalous sounds (or "electrophonic" sounds, as they could now safely be termed) – depended greatly upon the existence of factors in the nearby environment that were capable of acting as "receivers". A person with an Afro hairdo or wearing glasses might "hear" the sound whereas a bald-headed man with perfect eyesight standing nearby might hear nothing untoward. This explains the capriciousness of electrophonic sounds; one of the supposed strong points of the psychological model.

It should also be mentioned that the electrophonic effects are not only restricted to sound. A 1992 fireball in Oregon was accompanied by a whole range of electrophonic sounds, even causing a metal lamp in one house to emit a sizzling noise for a couple of seconds. It was also blamed for a motorist receiving an electric shock! Another witness reported feeling a pressure in his chest at the time of the fireball sighting, which may have been something more than a psychological/physiological reaction to the awesome experience.

By the way, electrophonic sounds are not confined to meteors. It has long been known that many people claim to "hear" auroral displays. This has also suffered from psychological explanations. Bright and active auroral displays frequently resemble waving curtains, and the suggestion was that the mind "heard" the swish and rustle of these cosmic draperies!

More commonly, lightning has been reported to "sizzle" (the writer has personally experienced this). Similar to the meteor events, this sound occurs simultaneously with the flash, long before the sound of thunder arrives.

Lightning also appears to display the same capricious nature as meteor electrophonics. For example, one instance involved a group of three people standing together, only one of whom heard a lighting flash "click." Recalling Keay's long-haired subjects, it is interesting to note that the person hearing the flash was a lady sporting a full head of thick hair, while her two companions were gentlemen with nearly bald heads!

Although the "Keay effect," as it is now known, is no longer controversial, mysteries remain.

For instance, whereas the mechanism readily explains electrophonic sounds associated with very bright meteors, anomalous sounds are also reported from time to time for much fainter meteors. This is something requiring further research.

Also, the theory indicates that electrophonic sounds are more likely to be heard when the meteor enters our atmosphere at a shallow angle. Yet, witnesses of the Karoonda meteorite, mentioned earlier, describe this object's approach as very steep, about 70° to the horizontal. As we have seen, several people reported hearing anomalous sounds as the fireball descended. (But remember the earlier conclusion of Ivanov and Medvedev concerning meteors arriving along a steep trajectory).

A bright meteor enters the atmosphere on a shallow trajectory. NASA image.

Another problem is the apparent directionality of at least some of the reported sounds. If the sounds arrive from "receivers" in the immediate environment, there is no reason to suppose that the direction from which the sound appears to come should coincide with that of the fireball itself. Yet, this is precisely what has been reported. The Murchison Meteorite witness, of whom we spoke earlier, looked upward and to the south as soon as she heard the sound, and immediately spied the fireball.

The directional aspect is even more pronounced in the following account of a 1989 fireball seen from rural New South Wales:

> [She] heard "what sounded like a heavily laded semi-trailer coming to the farm." [Sitting up in bed] she … then saw the fireball appear in the window. "The noise seemed to follow the fireball across the sky and sounded like an engine turning over."

Or, this account of a Russian fireball in 1990:

> Hear crackle and hissing like sounds of firewood burning, intensity about third that of an electric shaver. Sound came from the fireball during the whole flight and caused feeling of anxiety and uneasiness.

These are clear statements of directionality, although the fact that the first was witnessed indoors may be important. Perhaps something near the window acted as the "receiver."

Yet another unsolved issue is whether different types of meteoroids give rise to different types of electrophonic sound. After examining hundreds of reports of anomalous meteor sounds, Keay found that they could be grouped into three distinct categories, namely, sharp, staccato, and smooth.

The first include sounds classified as pops, cracks, bangs, and booms – the latter, by the way, *not* including the sonic booms that are heard after the fireball has disappeared. These frequently coincide with explosive fragmentation of the fireball and can occur when the meteor is not accompanied by any continuous electrophonic sound. Sounds of this class may accompany fainter fireballs that suddenly burst with explosive violence.

The staccato classification includes crackling, sizzling, and rustling sounds, while the smooth category are those described by such adjectives as "hissing," "swishing," "rushing," and "roaring."

Clearly, this work is still in its infancy, and there are sure to be many more surprises in store. What we can say, however, is that this investigation, spearheaded by Colin Keay into an interesting example of meteor "weirdness," has blossomed into nothing less than a new branch of science, the science of geophysical electrophonics, now set firmly on its foundations of extensive observation and laboratory experimentation. Gone are the psychological red herrings and excessive skepticism of reports that "didn't fit" existing knowledge.

The lesson worth learning from this is that once these "weird" reports were taken seriously, existing knowledge itself was expanded and new processes discovered. Who knows where this emerging science will lead in the years ahead?

PROJECT 16
Meteor Sounds

Fortunately, we can include electrophonic sounds in fireball reports these days without risking being called a crackpot!

If you are fortunate in seeing a bright meteor (especially one large enough to drop meteorites) take careful note of any sound heard simultaneously with the meteor's visibility. Note especially whether the sound (if one is heard) can best be described by any of the categories mentioned in the text ("hiss," "roar," "pop," etc.) and whether the sound gave the impression of following the meteor across the sky, whether it seemed to emanate from a particular direction on the ground, or whether it seemed to surround you – coming from no direction in particular. A careful description of what you hear may hold very important clues to the further study of this phenomenon.

5. Strange Stars and Star-Like Objects

When we think about for a few moments, stars are pretty weird things at the best of times! A ball of compressed gas heated to fierce temperatures by a continuously exploding thermonuclear bomb on steroids at its core may seem quite an odd sort of thing to have as the basic unit of the cosmos. Yet, if such objects did not exist, the cosmos would be plunged into an eternal night of utter lifelessness. Stars may seem a bit weird, but without them, the universe would be a pretty dull place. Not that there would be anyone here to complain about it!

From time to time, though, something more than the usual strange is reported having been seen among the stars. Sometimes this is a signal for an important discovery. At other times, it is a red herring that can leave the hapless astronomer with a similarly colored face!

Let's look at some instances of both classes of weird reports.

The Aries Flasher

In the wee small hours of the morning of September 1, 1984, veteran meteor observers Bill Katz, Bruce Waters, and Kai Millyard were engaged in their favorite occupation by the shores of Canada's Lake Huron when suddenly, to quote Katz, "a big flash went off west of the Pleiades." Assuming that a rare head-on or "point" meteor had just been witnessed, the trio duly recorded the event and continued their meteor vigil.

Point meteors are seen from time to time, but they are rare. Yet the event just witnessed aroused a sense of déjà vu. Just three weeks earlier, these same three meteor observers had witnessed a similar event, also near the Pleiades ("above the Pleiades" as they then described its position). The chances of the same three people seeing two point meteors just 3 weeks apart in the same small region of sky (maybe even at the same exact spot!) would be – pardon the pun – astronomically small!

But worse was to follow.

Their curiosity now well and truly whetted, the observers checked back through their earlier records and found that they had already recorded two other point meteors in the same general region during the previous year!

Clearly, something was amiss. The probability of recording so many head-on meteors from the same region in such a relatively short time period was just too slight to be due to chance. Something other than meteors had to be flashing in the Pleiades region.

The trio decided to give this part of the sky special attention, and as a result of this increased vigilance, a remarkable tally of five further flashes were recorded during the following three months. The individual flashes appeared as star-like points of light between zero and third magnitudes in brightness but lasted less than a second. Because of their very short duration, it was virtually impossible to pin down their position with any accuracy. All that could be determined was that the flashes were near the Pleiades, though apparently on the Aries side of the Taurus/Aries constellation boundary.

Katz and his colleagues had by then abandoned the point-meteor explanation. The number of events was simply too large and the time interval too short for the flashes to be chance sightings of head-on meteors.

Could these three meteor watchers have chanced upon some new type of bright optical burstar?

Now, what exactly is a burstar?

Initially detected as gamma ray bursts, these events were discovered during surveys of the sky at gamma ray wavelengths. It soon became apparent that there were two quite distinct classes – one class with bursts of longer than 2 s duration and a second group having shorter bursts.

The latter are now thought to arise from mergers of very compact objects such as neutron stars or black holes. At least some of these may be in our own galaxy.

By contrast, bursts of longer duration lie well beyond our home galaxy and are believed to be caused by the collapse of very massive stars. At least, that is one popular and credible-sounding hypothesis.

When stars having masses typically three to six times greater than the Sun collapse, their cores are crushed into neutron stars. The collapse itself sets off a spectacular stellar explosion known as a Type II supernova. According to the above-mentioned burstar scenario, when stars of even greater mass collapse, the core is compressed beyond the neutron star phase and becomes a black hole. In this instance, a supernova does not eventuate. (A "rebound" is necessary to trigger this, and no such "rebound" occurs if the core collapses into a black hole.) Nevertheless, the newly forming black hole swallows up the outer layers of the collapsing star, whirling it in a vortex at relativistic velocities (i.e., at speeds approaching that of light itself!) in the process. This creates a spectacular fireball of even greater brilliance than a supernova. The "failed" supernova ends up outshining the "successful" one by a factor of 100!

These brief but incredibly energetic events are more properly called "hypernovae." They have been observed in visual light and as far as we are aware, are the most energetic events to have occurred since the Big Bang. It is widely thought that most of the radiation is confined to quite narrow beams shining out from the core of the vortex and that the event is observed if one of these opposing beams happens to be pointed in our direction.

By any estimate, these are very dramatic events.

That the burstar explanation for the flashes was the thought initially uppermost in the minds of the discoverers is reflected in the unofficial name OGRE (Optical Gamma Ray Emitter) given to the object in early accounts. However, neither of the above-mentioned explanations seem capable of explaining repeated bursts, but at the time of these reports we knew even less about burstars than we do today, and the thought of a repeating burstar giving very bright optical flashes appeared more credible than it may today.

Nevertheless, there were other more obvious problems. No gamma ray bursts had been observed from the region, and the estimate of the flashes position(s) was still not so accurate as to determine whether they were coming from the same *exact* location or merely from the same region of sky. For a burstar to be the explanation, all the flashes would, of course, need to appear at *exactly* the same point.

In the hope of pinning down the position of the "Aries Flasher" (as it had become more popularly known), photographic patrols of the region were brought into action. On March 18 of the following year, these patrols bore fruit. On that night, a star-like point blinked in and shone brightly for just quarter of a second, before blinking out again as completely as if it had never been. It was brilliant, estimated as magnitude –1 or intermediate between the brightness of Sirius and Canopus, the two brightest stars in the night-time sky. Had the Aries Flasher finally been caught on film?

There was a problem, however. This flash was not in Aries at all but in the neighboring constellation of Perseus!.

Indeed, by the time the photograph had been secured, an ever increasing number of star-like flashes were being visually reported from the general region of Aries, but an annoying fact was also starting to emerge. No two flashes had exactly matching positions. True, positional measurements were, by the nature of the events, only approximate, but by the end of 1985 the discrepancies had grown to around 6° – that's equivalent to 12 breadths of the full Moon! There is approximate and then there is *approximate*. When we are dealing with experienced observers (such as the initial trio and many other astronomers who witnessed "The Flasher"), it is not easy to fall back on the excuse of observational error when angular distances of this order are encountered.

By the end of 1985, it began to seem as though anything that flashed, glimmered, twinkled, or blinked anywhere between Pisces and Andromeda was instantly hailed as another sighting of the Aries Flasher. Yet, as more observations were accumulated, the less accurately could the phenomenon be pinned down. That is not the way things are supposed to work!

Throughout this saga, the Aries Flasher had its skeptics.

For instance, Rob McNaught of Siding Spring Observatory was never a fan of the Flasher. The whole thing, according to Rob, was just a flash in the pan that would soon pass away into history. He argued that the Flasher was nothing more substantial than glints of sunlight from artificial satellites. This would not only account excellently for the brevity and considerable brightness of the flashes, but it nicely explained why no two people saw the Flasher in exactly the same place. Of course, flashes from satellites take place in other regions of the sky, but once the legend of the Aries Flasher became established, any that happened to occur in a broad region around that constellation were afforded special significance.

Independently, Paul Maley reached a similar conclusion.

As time passed and the Flasher refused to be pinned down or confirmed by corroborating observations of gamma ray bursts, the Maley/McNaught explanation became generally accepted and the issue was quietly dropped.

Glints from artificial satellites can be quite startling actually, and it is not surprising that random sightings of these events have been misinterpreted as something unusual. The present writer recalls an incident in 2001, while awaiting the Leonid meteor storm of that year, when something like a flash bulb exploded near the zenith. At first, I thought I had seen a head-on meteor, but a flash from a satellite was far more likely. That flash was over in a split second, but on another occasion, I saw a satellite "flare" from below naked-eye visibility to a shadow-casting brilliancy at least as great as Venus, only to drop back to obscurity in a matter of 2 or 3 s. Satellite flashes had become quite common by the time of this event; otherwise I can't imagine what I would have thought that I was seeing!

ABBA I

While on the subject of flashing satellites, we should mention (though with hushed voices) what must surely have been one of the most absurd reports ever received by an astronomical organization. The report appeared in the electronic circulars sent to subscribers of *The Astronomer* magazine, albeit with the note from the editor that "There is no reason to believe that this claim is true and this should not therefore be circulated outside our group." This was followed by the request, "Would someone like to check the above figures (i.e., the elements of a supposed orbit – see below) and e-mail comments?" Rob McNaught himself furnished a further refutation of the whole incident which, in common with the Aries

Flasher, he adequately explained in terms of flashing satellites. But, whereas the Aries Flasher incident was an understandable mistake, the kindest thing that could be said about the Abba I report is that it could only have been true in an alternative universe!

So what was Abba I?

Would you believe, a flare star following an elliptical orbit *within the inner Solar System*?!

The announcement was made on May 10, 1988, by someone in Arizona who witnessed a series of flashes on February 16, March 19, and April 29. The initial report had been sent to another observer in the United States, who in turn forwarded it on to *The Astronomer*. Assuming that the flashing object was orbiting the Sun (though why that assumption should have been made is unknown), the discoverers calculated an orbit confined to the inner Solar System and having a period of just 1.23 years!

Of course, random flashes from artificial satellites in *terrestrial* (not *solar*) orbits are entirely adequate to explain these observations. It is hard to believe that the persons reporting these events were unaware of this. Surely they were having a joke (a belated April Fool's Day joke, perhaps?), and sat back laughing at the reaction. Was the name given for the "discoverer" even real? Yet, the report was made *as if* seriously intended and had to be treated accordingly.

Of course, there is no possibility that a flare star exists within the inner Solar System. Flare stars are actually red dwarfs – small stars giving out less than one thousandth the light of the Sun. Paradoxically, these comparative midgets experience extremely strong flares and, because of their feeble light, these actually increase the stars' total brightness noticeably for several minutes before they return to normal. Our own Sun has flares, but because they are milder and the Sun's total light greater, they contribute negligibly to the Sun's overall brightness.

Nevertheless, even though a flare star is faint when compared with the Sun, it is still a star! It is still a very hot and (in comparison with Earth) large body whose presence within the Sun's planetary system would certainly not have gone unnoticed. Our sky would have a second Sun – a smaller and dimmer one, admittedly, but a Sun nevertheless. Moreover, a flare star in the orbit computed for Abba I would periodically pass quite close to Earth, where it would blaze brilliantly as well as raise the temperature of Earth to fatally high levels. Not that we would be around to worry about this, however. The gravity of such a large body would either have flung our planet into interstellar space or into the Sun long before the human race appeared on its surface!

This is all that needs to be said about Abba I. Probably even mentioning it is saying too much, but perhaps it deserves at least passing mention in a book about weird astronomy, though surely the weirdest thing here is the report itself rather than the supposed object being reported!

Dr. Hafner's Blinking Star

It is with a sense of apology that we include the following story in the same section as Abba I. Unlike this, the following event proved to be a real phenomenon that aided in the advance of our knowledge of a rather rare type of stellar system. Not all flashing or blinking stars lead to dead ends!

On the night of July 1–2, 1988, Dr. Reinhold Hafner, then a visiting astronomer at the ESO LaSilla Observatory, must have thought that his eyes were playing tricks when one of the faint stars on the screen in front of him suddenly disappeared. No change occurred in any of the other stars in the field. Only this one simply vanished!

Dr. Hafner figured that an unseen and much fainter companion must have eclipsed the star and that eventually it would reappear as the eclipsing body moved out of the line of sight. How fortunate to have observed such an interesting event!

Imagine his surprise when, just a few minutes later, the star popped back into view again!

Dr. Hafner's interpretation of the odd event had been correct. The star was indeed an eclipsing binary, i.e., a star with a fainter companion that periodically crosses our line of sight and hides it from view. But this one was of a far more extreme nature than he had dared to imagine. In fact, what he had just discovered was the most complete and probably the faintest stellar eclipse ever seen.

Two further eclipses were observed during the same and the following night, and, after the data was thoroughly analyzed, the true nature of the object became clear.

The star itself is about 25,000 times fainter than the faintest discernible with unaided eyes on a dark night, and lies in the constellation of Ophiuchus. Known rather unromantically by its catalog designation as PG 1550+131, it was first observed at Palomar Mountain Observatory in the mid-1970s, but its eclipses were not noticed then. What was discovered at that time was its unusual blue color. Somewhat later, it was also found to be slightly variable. This variability was a real feature of the star itself, in no way associated with the "blinking out" of the eclipses.

The unusual color plus its variability and, of course, its extreme eclipses led Dr. Hafner to take a closer look at this stellar oddity. What he found was truly fascinating.

His strange blinking star was an example of a relatively rare type of object known as a pre-cataclysmic binary. So what is a pre-cataclysmic binary?

Well, a binary star is a pair of stars that orbit a common center of gravity. In a minority of binary star systems, one of the components has already passed through its entire evolutionary sequence as an active hydrogen-fusing star and collapsed into a small and extremely dense object known as a white dwarf. The other star in these types of systems, however, is still in the hydrogen fusing, or "main sequence," phase of its evolution and, compared with the white dwarf, remains a relatively low-density object. The two stars remain so closely bound to each other that the strong gravitational pull of the collapsed member of the pair teases out a constant plume of gas from its companion. This gas eventually spirals down onto the surface of the white dwarf, where it steadily accumulates until such time as it becomes unstable and explodes in a sudden burst of brilliance. The resulting temporary surge in the star's brightness can be dramatic, but it is not destroyed by the eruption. In fact, it is not affected very much at all, and once the event is over, the stream of material begins accumulating again in readiness for the next blast. Binary stars of this type are called "cataclysmic variables."

A pre-cataclysmic binary is a potentially cataclysmic variable star that has not yet "gone off." In these events, the gas stream

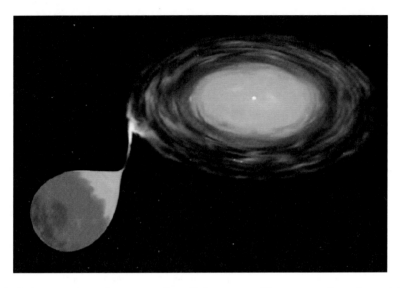

Artist's impression of an accretion disk surrounding a cataclysmic variable star. Credit NASA/STSci.

from the less dense component has yet to build up to the degree where an explosion can occur.

Although a relatively large number of cataclysmic variables have been observed, only a handful of stars in the immediate pre-outburst state are known. Hafner's peculiar blinking star is one of them.

The extreme nature of the eclipses of PG1550+131 sets clear limits on the physical properties of this system – for example the size and shape of the orbits of the component stars, their respective temperatures, and the like. Determining these values provided scientists with an excellent look into the anatomy of one of these rare star systems and yielded valuable information, not just on this system but on pre-cataclysmics in general.

The eclipse obviously took place when the fainter star passed in front of the brighter one and amounted to a record fading of nearly a hundred fold. This meant that the fainter star is about a hundred times dimmer than the brighter one. Moreover, the unprecedented short duration of the eclipse meant that the fainter star also has a very small diameter. Putting together its faint magnitude and small diameter, astronomers deduced that it must be a red dwarf; a cool star (about 5,400°F or 3,000°C, which is cool for a star!) near the lower end of the main sequence. These stars use up their hydrogen supplies slowly and, though faint, they are very long-lived.

The brighter star is the collapsed component – a white dwarf with a surface temperature around 32,000°F (18,000°C). In its youth, this star was further up the main sequence than its diminutive companion. That is to say, it was brighter, hotter, and a more vigorous consumer of its stock of hydrogen. At that time, it would have far outshone its red dwarf companion, but the price it paid for this comparative extravagance was a rapid and (from the red dwarf's point of view) premature aging. This is why we now see this very evolved (or should we say "decayed"?) star in tandem with its small main-sequence sister.

The two stars orbit the system's center of gravity every 187 min, and the distance between the two components is, in round figures, a mere 438,000 miles (700,000 km). The entire system could fit within our own Sun!

Because the two stars are so close together, the hotter white dwarf actually heats the facing hemisphere of its cooler companion to nearly 11,000°F (about 6,000°C), twice the value of its "normal" temperature!

Surely, Dr. Hafner's accidental discovery of this remarkable object was one of the happier instances of "blinking" stars!

Ejnar Hertzsprung's Enigma

Ejnar Hertzsprung was one of the leading figures in twentieth century astronomy. He is immortalized in the Hertzsprung-Russell diagram, which classifies stars into their various types. What astronomy student has not had to memorize the spectral types along the stellar main sequence of the H-R diagram?

Well, when somebody of Hertzsprung's caliber comes up with a mysterious object, it is not so easy for the skeptic to discredit the observation.

The object was not actually seen by Hertzsprung. In fact, it was not really *seen* by anyone, but it *was* photographed. On December 15, 1900, an image was left on two Harvard Observatory

"Cataclysmic Dawn"; artist's impression of an early morning scene through the mouth of a cave on a hypothetical watery planet orbiting a cataclysmic variable star. The artist writes that this scene is unlikely to be found anywhere in the real universe, as such planets are not likely to exist in these environments. But it does make for a spectacular scene! © Mark A. Garlic 1996.

photographs taken one hour apart. There the photos lay for almost 27 years, unseen in the archives.

Hertzsprung knew nothing of the image before coming across it on April 1, 1927 (no significance should be placed on the date!). He discovered it while meticulously examining thousands of photographs in the Harvard collection for evidence of variable stars. The object he found was clearly variable, but it did not really look like a star!

The mystery object was quite bright. From the photograph, it seemed bright enough to be just within naked-eye range for a keen-sighted observer under an excellent sky. It also appeared to have a definite size, unlike the point source of a star, and (even more puzzling) appeared to have increased in diameter during the hour between the first and second photograph.

Had there been just one photograph, Hertzsprung would have dismissed the image quickly as a photographic plate flaw. However, the chance that two similar flaws should appear in consecutive photographs of the same region of sky *at exactly the same position* seemed to him too much of a coincidence to be credible.

The problem for Hertzsprung was that nothing else seemed credible either!

The rapid variability suggested that the object was not at stellar distances (though discoveries since his day show that this assumption was not necessarily true), and he therefore suspected that it lay within the Solar System. This immediately suggested a comet; however, he ruled out this explanation because plates taken of adjacent regions both prior to and following the date in question did not show any similar object. Only a fast-moving comet could have avoided being imaged in the other fields; however, the apparent lack of movement during the hour between photographs ruled out that possibility. If the object was moving, it must have been drifting only very slowly against the background of stars – too slow for its motion to be perceptible during the course of an hour.

Hertzsprung even briefly entertained the possibility of it being a cloud of debris released from an asteroid collision, but he decided that the image was too round and regular to result from such a cataclysm.

Although Hertzsprung did not think that his mystery image was that of a variable star, others were not so sure. Richard Prager actually listed it as Number 122 in his 1934 catalog of suspected variables, and prominent husband and wife astronomical team Sergei Gaposchkin and Cecilia Payne-Gaposchkin suggested, in their 1938 monograph on variable stars, that the object may have been an example of a hitherto unrecognized class of very rapid novae.

Then, in 1951, variable star expert Dorrit Hoffleit suggested that a flare star of unusually large magnitude range might have been responsible for the mystery images. Hoffleit speculated that the very red light of the flare star may have been responsible for the image appearing "nebulous" instead of stellar on the type of photographic plates being used in 1900.

Unfortunately, there is no evidence for any of these suggested explanations.

A different tack was taken by amateur astronomer Thomas Anderson soon after Hertzsprung's discovery was announced. In a letter to Hertzsprung dated May 10, 1927, Anderson revived the comet suggestion, though with a difference. He suggested that the object may have been a normally faint comet caught during the process of a brief but strong brightness flare, similar to that suffered by Holmes' Comet in 1892. This latter comet brightened from obscurity to naked-eye visibility within a matter of hours and displayed a remarkably symmetrical appearance not unlike the images found by Hertzsprung. (The Holmes performance was, by the way, repeated in even more spectacular fashion late in 2007 … but that is another story!).

Hertzsprung liked this suggestion and requested that the Harvard staff carefully examine the plates taken on the nights before and after December 15, 1901, for further images of the purported comet. Unfortunately, nothing was found.

The absence of any suspicious image on plates taken immediately after December 15 does not bode well for Anderson's explanation. A comet such as Holmes might have brightened fast enough to have passed from obscurity to a relatively bright object in a single day, but it is unlikely to have faded with equal haste. Holmes' Comet itself remained bright for several weeks, and other comets that have experienced major flares have invariably brightened a lot faster than their subsequent fading. Although some *have* faded faster than others, this fading has still not been fast enough to explain the absence of Hertzsprung's object on the nights immediately following December 15. Admittedly, some comets have been seen to experience minor flares lasting only for one day or thereabouts, but these flares have all been far too small in amplitude to account for the rapid disappearance of something as bright as Hertzsprung's mysterious phenomenon. The comet – if that is what it was – would have been bright enough both prior to and following such a minor flare to have left an image on earlier and later photographs.

Moreover, the apparent expansion of the object during the course of just one hour makes the Anderson explanation suspect. Rapid expansion does take place during a major flare, but conspicuous enlargement within so short a space of time would be unprecedented.

So the question remains. What *was* Hertzsprung's mystery object?

It is easier to say what it was not, or at least, what it *probably* was not! For a start, it was *definitely not* a head-on meteor. This potential explanation is immediately precluded by the length of time between photographs and was not suggested by anyone.

Was it a very fast nova of some other type of variable star?

Unlikely. There is no evidence that novae rising and falling fast enough to explain the mystery even exist. Flare stars exist and certainly can perform quickly enough, but no suitable star is known at the position of the mystery object, and it is far from certain that one could give rise to the sort of image recorded.

The present writer even briefly entertained the thought of an optical counterpart of a very large gamma-ray burst, something completely unknown at the time when Hertzsprung found the image, let alone in 1901. Nevertheless, the same problems with the image that appear to wipe out the flare star suggestion hold here as well, in addition to the comparatively long duration of the event.

A comet (maybe in outburst) would certainly explain the diffuse appearance of the image, but it is difficult to explain why it was not also recorded on the previous night and subsequent nights. A sudden flare might explain its absence on previous nights, but, realistically, it should still have been captured on the Harvard plates taken after December 15.

A very interesting possibility is that the image represents some rare event that has not even been recognized as yet. Remember the gamma-ray flare suggestion and the earlier remark that such a phenomenon was not even known in 1901? Indeed, the known physics of that time did not even allow for the existence of such a thing. Could Hertzsprung's object be an example of some phenomenon still unknown at the present day?

That would be the most interesting possibility, but in all honesty, we must admit that it is a long shot. Well over a century has rolled past since the image appeared on those photographic plates, and the very fact that we are still puzzling over it is proof enough that nothing closely similar has been seen in all those years.

From the exciting, albeit very remote, possibility of an unknown phenomenon, we must now go to the other end of the spectrum and entertain the most mundane of possibilities. Perhaps this was not a real object after all. Maybe the mystery is no more than a very unusual coincidence of two plate flaws appearing at the same spot on two consecutive photographs. Hertzsprung thought this unlikely, but if nothing else explains it, the unlikely becomes the last resort!

This explanation might not be very exciting astronomically, but such an improbable happening would be statistically interesting in its own right. The probability of it happening may appear *so* low that not many people will be satisfied with this suggestion, but extremely low-probability events do happen from time to time. During the World War II bombing of London, to recall just one example, a bomb fell through the roof of a house and failed to explode. Not long thereafter, a second bomb fell through the hole made by the first ... and likewise failed to explode! What are the chances of that happening? Could it be even less probable than the Hertzsprung object simply being two plate flaws?

The Spooky Star of Halloween

Sometimes an announcement of an unusual discovery comes at a very odd time. One example of this was the announcement in

1986 that very complex organic material had been discovered in Halley's Comet. This information was released on April 1! Apparently, when it was posted on the notice board at Siding Spring Observatory, one skeptical astronomer appended a note "Is this an April Fool's Joke?" underneath. As it turned out, the answer was "No!"

Something similar happened on October 31, 2006. On that traditionally spooky night, an announcement came over the wire that Akihiko Tago had noted a mysterious brightening in an otherwise very ordinary and inconspicuous star in the constellation of Cassiopeia. Of course, there is nothing very odd about variable stars, but in a star that had given no prior hint of instability, nor which belonged to a potentially variable class of star, this sudden brightness jump of 50-fold *was* odd – very odd indeed! From needing a small telescope to even glimpse, the star suddenly popped almost into the range of opera glasses.

Needless to say, some astronomers thought that they were being made the brunt of a Halloween prank. But a second e-mailed message from the Central Bureau of Astronomical Telegrams convinced them that this was treat rather than trick. Something both odd and interesting was happening, and astronomers all over the world quickly sprang into action.

Among the early observers of this strange star were several members of the Center for Backyard Astrophysics network, founded by David Skillman and Joseph Patterson. Two CBA members in particular, Robert Koff in Colorado and Thomas Krajci in New Mexico, made very valuable observations of the star. They were able to further confirm that the star was not really weird at all – at least, it should *not* have been weird. Everything pointed to it being a perfectly normal main-sequence star of the A type. In other words, it was similar to the well known Vega, only 130 times more distant. Stars of that class simply do not vary in brightness!

Not only did the star rise suddenly, but the observations of Koff and Krajci showed that it also faded fast after reaching peak brightness. Other observations revealed no sign of unusual emission lines in the spectrum, no change in color, no detectable X-ray emission, and (from studying survey plates taken between 1964 and 1994) no previous indication of variability.

So what really did happen on Halloween night 2006?

Incredible though it seems, when the behavior of the star was graphed, it looked spookily similar to the light curves of micro-lensed stars, except that it was much, *much* brighter than any microlensing event ever seen.

Microlensing is a subspecies of gravitational lensing, a phenomenon predicted by general relativity theory. According to this theory, what we call "gravity" is essentially the warping of space itself in the proximity of a massive body. Light, as it is transmitted through space, follows this warp, as was famously shown by the displacement of star images close to the Sun during the total solar eclipse of 1919. Counter intuitively, relativity theory predicts that when a sufficiently massive body passes in front of a distant background luminous object, the warped space around the "eclipsing" object acts as a lens, and, instead of being hidden from view, the light from the point-like background one is focused into multiple images. This strange phenomenon is pre-dicted by General Relativity but (as we shall shortly see) was not taken seriously by Einstein himself.

Gravitational lensing was first observed in quasars. In 1979, what appeared to be two quasars flanking a single galaxy was shown to be a pair of images of a single quasar gravitationally lensed by the (far closer) foreground galaxy. Then, in 1986, Bohdan Paczynski of Princeton pointed out that individual stars should also act as lenses focusing the light of more distant ones. However, because of the weaker gravity of an individual star, the multiple images should be separated by less than 0.001 s of arc, too small a separation to be detectable with present-day tech-nology. Nevertheless, such *microlensing* events should still be detectable through an amplification of the light of the more dis-tant star, and this amplification should present a recognizable profile, distinguishing microlensing events from intrinsic vari-ability of the star itself. By monitoring very rich star fields (e.g., toward the center of the galaxy or in the neighboring galaxies known as the Magellanic Clouds), many such events have now been recorded, and it has even been possible, by studying the exact profile of the more distant star's light variation, to show that a few of the lensing stars have planets orbiting them. But because this geometric alignment has such a low probability, most of the events thus far monitored have been very distant and very faint.

At least, until Halloween night 2006!

If this interpretation of the 2006 event is correct, it seems that a faint star passed in the line of sight with the bright but relatively distant A-type one and focused its light in our direction. Such an event involving a relatively bright star is estimated to occur once every 30 years or thereabouts, so Tago's chance discovery was truly a remarkable feat.

By the way, microlensing has an interesting, if rather short, history.

As mentioned earlier, although the phenomenon is a direct consequence of General Relativity, Einstein himself was very dismissive toward it.

The great scientist, despite his brilliant insight into the nature of physics and his radical political ideas about globalism and pacifism, remained surprisingly conservative in many respects. The story of his resistance to quantum theory is well known, as is his refusal to believe his math when this pointed toward a non-static universe (the famous "cosmological constant blunder"). But his lesser-known attitude toward microlensing betrays an equally cautionary streak.

In 1936, Einstein wrote a paper in which, while admitting the existence of the microlensing effect, famously stated that it was unlikely that this phenomenon would ever be observed. It is actually quite likely that he worked out the theory of microlensing as early as 1912, but set it aside as being of little interest!

It seems that the 1936 paper was not Einstein's idea at all and the phenomenon of gravitational lensing would not have been brought to the notice of the world had it not been for a certain Rudi Mandl, Hungarian engineer and amateur at physics.

Mandl wrote to Einstein several times pointing out this consequence of his theory, but his letters went unanswered. Apparently Einstein dismissed him as being a bit of a crackpot!

Undaunted, the indefatigable Mandl then traveled all the way to Princeton to put his case to Einstein face to face. Finally, Einstein gave in and wrote the paper, but not with good grace, it seems. Following the paper with a rather jaundiced letter to the journal *Science*, he wrote that the result was a useless one that he only made known because "Mister Mandl squeezed [it] out of me" and that he published it because "it makes the poor guy happy."

This is where the situation remained until 1979, when the first gravitationally lensed quasar was discovered. Following Paczynski's recognition of microlensing by individual stars, the subject became a hot one and developed an importance in the hunt for extraterrestrial planets that not even Mandl (let alone the skeptical Einstein!) would have dreamed possible.

Interestingly, microlenses have even been found within larger gravitational lensing events. On at least one occasion, a temporary brightening in a lensed quasar appears to have been due to secondary microlensing by a star in the *remote lensing galaxy!* In other words, as astronomers on Earth observed the very remote quasar shining through a (nearer but still remote) intervening galaxy, one of the stars within that galaxy passed directly in the line of sight of the quasar, creating its own microlensing event. By analyzing this secondary event, astronomers were even able to get some idea of the type of star involved. It turned out to be a red dwarf, in a galaxy far, far away!

By the way, a gravitational lens really produces four images of the lensed object, although in most instances only two are obvious. (We refer here to images of quasars lensed by intervening galaxies. As mentioned earlier, the individual imagers of microlensed stars cannot be distinguished by present-day means.) The highest resolution gravitational lenses, however, do display the four images, flanking the lensing galaxy like a cross. The best example of this is widely known as the Einstein Cross – a rather ironic title considering Einstein's skepticism. Perhaps it should be renamed the Mandl Cross in recognition of the unsung hero of gravitational lensing!

The Weird Flare of 2006

Have we discovered all the different types of objects that populate our universe?

It might seem a bit arrogant to think that we have, but the list of known object types is nevertheless impressive. We have galaxies, clusters of galaxies, quasars, stars, nebulae, planets (giant, megagiant, large, and small), brown dwarfs, white dwarfs, black holes, and so on. The real test of how well we have tagged the cosmic zoo is to see whether or not all observations can be accounted for by one of these varieties. If everything we see fits into one or another of the categories of known objects, then chances are we have discovered at least all the broad classes of astronomical beasts.

Nevertheless, if there is one task harder than proving that there is a needle in a haystack, it is proving that there is *not* one there. Finding a needle in a haystack may be notoriously difficult, but if it is accomplished, it becomes its own positive proof. But not finding one can mean either that one is not really there *or* that one is there but we have simply not located it as yet. Likewise, not finding an observation that cannot be explained by evoking any known category of object may just mean that there are some very, very rare objects lurking out there that have thus far eluded us. It is with this in mind that the events of 2006 take on their importance.

That year, the Hubble Space Telescope recorded something that does not fit with – or is very difficult to fit with – the behavior of any known type of object. Perhaps something truly new has come into our vision.

The object was first recorded by the Hubble on February 21, 2006, as a dim pinpoint of light in the constellation of Bootes. For the next 100 days, its brightness rose steadily, reached its peak, and thereafter faded to oblivion during the course of a further 100 days.

Nothing – no faint star, galaxy, or nebula – could be found at the object's position once it had faded from view, although an X-ray source was apparently detected at the relevant position during the declining phase of its visibility.

The rise and fall of the object's light was unprecedented in the history of astronomy. Supernovae take no more than 70 days to rise to maximum, and their subsequent behavior is very different from that of this object. Nova, gravitational lensing, optical counterparts of gamma ray bursts, and the like are all much faster.

Moreover, the spectrum of the mystery object (now cataloged as SCP 06F6) is made up of lines that do not appear to match any

November 2005 - January 2006

May 21, 2006

Optical Transient SCP 06F6
Hubble Space Telescope ▪ ACS/WFC

NASA, ESA, and K. Barbary (University of California, Berkeley)   STScI-PRC09-04

Hubble image of the strange flaring object of February 21, 2006. NASA/STScI.

known element, although it has been suggested that they may be carbon lines strongly shifted toward the red end of the spectrum. If that is true, the object must be racing away from us at tremendous speed. It is difficult to understand how it could be accelerated to this velocity if it is relatively local. (Still, as we do not know what it is, we cannot be too sure about this, either!). On the other hand, if the red shift is cosmological, i.e., due to the expansion of the universe, the object must be about 1 or even 2 billion light years distant. If it is some sort of star, its brightness during the flare would then have been truly incredible.

Suggestions as to the nature of the mystery flaring object include the core collapse and explosion of a carbon rich star, a collision between a white dwarf (collapsed star) and either an asteroid or a black hole, the disruption of a star during an encounter with a black hole, or maybe yet another type of supernova. None of these has yet emerged as a clear winner, however, and the possibility remains that the real explanation lies in the discovery of something of which we simply have no knowledge. Further detection of similar events (should any occur) may eventually hold the key to the mystery.

Those Popular Pleiades

What is it about the Pleiades that has ingrained them so deeply into the human psyche?

Certainly, this cute little cluster of reasonably bright stars in Taurus makes a pretty sight, but why this grouping of stars should rate so highly in the folklore of peoples scattered across the face of the globe is unclear. The cluster is mentioned in the Bible (Job 38:31) "Canst thou bind the sweet influences of the Pleiades?" God asks Job. Far from Job's Middle East, the Australian

The Pleiades star cluster. The dust cloud through which the cluster is passing reflects the light of its stars. NASA/Courtesy nasaimages.org.

aborigines (Koories) hold a very similar notion as to the "sweet influences" of the cluster. This is even more remarkable when one considers that Job and the Koori people lived in two different hemispheres, so explanations involving climatic conditions at the time the cluster first appears in the dawn sky, is prominent in the evening – or whatever – do not hold. The Kooris say that the cluster is "very good to the black fella" and hold a New Year's corroboree (ceremonial dance) in its honor.

Then, far to the east, the Japanese know the group as Subaru. Today, a stylized representation of the cluster is used as the symbol for a make of motor vehicle of that same name.

PROJECT 17
How Many Pleiades Can You See?

Most people of average eyesight can see six Pleiades without optical aid, but there are reports of considerably larger numbers having been detected, either by folk with abnormally acute sight or during exceptionally clear conditions or a combination of both. One especially keen-sighted observer managed to see as many as 18 members of the cluster without resorting to optical aid!

How many can you see?

Choose a very clear night, a dark sky, and a time when the star cluster is at maximum altitude. Then remain in the dark for about 20 min, until your eyes are well and truly dark adapted before making your attempt. Make sure that no light source other than the stars is around to spoil your dark adaptation.

A good trick when seeking objects near the limit of visibility is to use averted vision; look just slightly to the side of the object being observed, while keeping your attention focused on it. By doing this, you allow light to fall on the more sensitive parts of the retina, and something that lurks just below the limit of direct vision may suddenly pop into view. The danger is that averted vision can become "averted imagination," but if the suspected star can be held in vision, it is almost certainly real. A check against a chart of the cluster, or a glance through small binoculars, will give final confirmation.

Among the ancient traditions of many nationalities are strange associations between the Pleiades and the Noah's flood as well as connections with celebrations of the dead. Festivals of the dead are linked with this star cluster in the religious rites of ancient Hindus, Egyptians, Persians, Peruvians, Mexicans, and Celts – a very mixed group, to say the least!

Equally noteworthy, many of these festivals are held in the month of November. In India, for instance, November is called the month of the Pleiades, and the 17th day of the month is a festival of the dead known as the Hindu Durga. Similarly, in Persia, the month of November was known as "Mordad," the angel of death. Far away in Peru, the same month saw the Feast of the Dead, also considered to be a New Year's festival, while in Ceylon, a combined festival of the dead and of agriculture takes place at the beginning of November.

Celtic religion apparently held the first day of November to be a night of mystery, when the reconstruction of the world was celebrated, and the tradition of All Saints' Day (and October 31 as All Hallows' – Saints' – Eve or Halloween) became part of the calendar of the Christian church.

All of these festivals appear to be associated with the time of year when the Pleiades cross the meridian at midnight.

Even in relatively modern times the Pleiades continued to cast their spell. In 1748, astronomer J. Bradley made the claim that the brightest star of the group – Alcyone – is the center of a system that includes our own Sun and its planets. The Solar System orbits it, according to Bradley. More recently, Sidney Collett took this a few steps further by claiming that the whole universe revolves around it! In support of this claim, he argued that "Pleiades" means "hinge or pivot" and "Alcyone" means "the center."

There is no possibility that this has any foundation in fact, of course. We now know that there is no gravitational center of the universe, and the Solar System itself orbits the hub of the Milky Way Galaxy. This galactic center is not the Pleiades, but a far larger and more remote system of stars in the constellation of Sagittarius.

Yet, even if this group of stars is not the hub of any larger system – and certainly not the hub of the universe itself – it remains the hub of many myths and stories. This, in itself, is a phenomenon worthy of further investigation.

Although hardly a profound mystery, a further weird coincidence concerning this star cluster needs a brief mention. On photographs, the cluster is seen to be entangled in a beautiful nebulosity, a fact that brings to mind the oft-quoted lines of Alfred Lord Tennyson's poem "Locksley Hall":

> Many a night I saw the Pleiades rising thro' the mellow shade
> Glitter like a swarm of fire-flies tangled in a silver braid.

But Tennyson knew nothing of the Pleiades nebulosity! His poem was composed in 1835 (although not published until 1842), whereas the first inkling that nebulosity cocoons the Pleiades did not come until 1859. That year E. W. L. Tempel found what he initially thought to be a large but very transparent comet close to the star Merope. When the "comet" refused to move, he concluded that what he had really found was a nebula, but not all of his fellow astronomers managed to see it, and its reality was not accepted beyond doubt by everyone. Others did detect it, however, and there was even some suspicion of a certain fuzziness around other of the cluster's stars. But it was not until the late 1880s that photographs revealed not only conclusive proof of the Merope nebula but also clear evidence that it was just one part of a wider system of nebulosity engulfing the entire Pleiades cluster. Then, in 1890, E. Barnard discovered an interesting nebulous knot within the wider nebula and very close to Merope itself. This is now known as "Barnard's Merope Nebula" or, more formally, as IC 349.

As astrophotography improved last century, so the real extent and beauty of the Pleiades' "silver braid" became apparent. Spectroscopy revealed the nebulosity to be comprised of dust, not gas, as initially thought. Without the cluster's stars, it would be invisible.

For a long time, the nebula was thought to be a remnant of the cloud from which the cluster formed. However, more recent work has determined the age of the stars to be greater than that implied by earlier estimates. They are simply too old to remain entangled in their birth cloud. Contrary to long held belief,

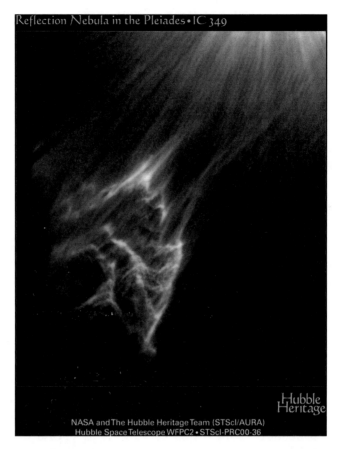

Reflection Nebula in the Pleiades • IC 349

Hubble
Heritage

NASA and The Hubble Heritage Team (STScI/AURA)
Hubble Space Telescope WFPC2 • STScI-PRC00-36

Hubble image of the bright nebulous knot known as Barnard's Merope Nebula. Merope lies just outside the image at upper right. The streaks extending toward upper right are composed of heavy particles falling toward the star. The bright nebulosity at center and lower left consists of fine particles repelled away from Merope by light pressure. NASA and The Hubble Heritage Team STScI/AURA.

the cluster's placental nebula has long since dispersed, and the beautiful "silver braid" now enveloping it is simply an unrelated cosmic dust cloud through which the stars just happen to be passing. Confirmation of this comes from precise measurements showing the motion of stars and cloud to be quite different from one another.

Although the Pleiades sans nebulosity is like Saturn without its rings, the fact is that the present photogenic appearance of the cluster is only a temporary phase and that one day the swarm of fireflies will again be untangled. We are simply fortunate to be around at the time of this chance cosmic encounter.

PROJECT 18
The Entangling Silver Braid

Although the true extent of the Pleiades nebulosity is only apparent on photographs, Tempel's nebula – the brightest region around the star Merope (this can be seen clearly in Fig. 5.4) – is visible in surprisingly small instruments. The writer detected it while using a pair of 20×65 binoculars, but the sky needs to be clear and dark for the very dim "stain" on the background sky to be noticed. Moreover, any dew on the lens can give a false reading, so be very skeptical if "nebulosities" appear near each of the Pleiades! But if you are satisfied that you are seeing the real thing, try progressively smaller apertures until it disappears. It is often quite amazing just how small an instrument is required to see traditionally "difficult" objects.

Another persistent Pleiades puzzle is that of the missing Pleiade. Most people count six obvious members of the cluster, though slightly better than average eyesight or very clear skies can bring one or more faint ones into marginal naked-eye visibility. Yet, the alternative name for the group is the "Seven Sisters." Why *seven*? Most people, surely, would call it the "Six Sisters."

Yet, the legend (?) that one of the Pleiades has disappeared is almost universal and very ancient. For example, the Bronze-Age Nebra Sky Disk, believed to have originated around 1,600 BC in what is now the German state of Saxony-Anhalt, includes a rather conspicuous group of seven stars that are thought to represent the Pleiades cluster.

Various explanations for the "missing Pleiade" have been proposed. In one version of the Seven Sisters story, one of the Pleiades wanders away from the cluster and turns into a comet.

The well known astronomy writer Mary Proctor suggested that the clear skies of ancient Syria permitted seven stars to be visible whereas the less favorable elevation of the group as seen

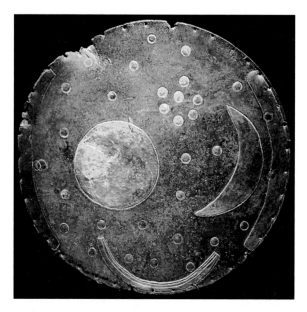

The Bronze Age Nebra Star Disk includes a group of seven stars, believed by most experts to represent the Pleiades cluster. Courtesy Wikipedia.

from ancient Greece allowed only six to be regularly observed. This explanation, while commendably simple, does not do justice to the widespread nature of the story.

Astronomer Cecilia Payne-Gaposchkin put forward a different suggestion. She pointed out that one of the fainter stars of the group, Pleione, sporadically blows what she refers to as "chromospheric bubbles," which may indicate a degree of instability. Maybe, she suggested, this star went through a bright phase around the time that human civilization first began to take notice of the stars, lifting it for a time from marginal to conspicuous naked-eye visibility.

We may also wonder if the "comet version" of the story gives a clue to the mystery. Could a bright comet passing through or near the group have sparked the legend? It is not likely that something as transitory as a comet could spark such a widespread legend, but, as an aside, it is interesting to note that Halley's Comet sometimes becomes visible close to this cluster as it heads inward toward the Sun.

Could the missing star have been a bright nova or supernova?

Such objects are unlikely to be found in a young cluster such as the Pleiades, but it is possible that a background nova or super-nova may have gone off and, for a while, appeared as an extra member of the group. Once again, though, the transitory nature of a nova meets the same problem as a comet (although it would at least stay in the same place for longer!). A supernova, on the other hand, would be brightly visible for months, but it would also have left an observable remnant, which should still be detectable today. Unfortunately, there are no supernova remnants beyond the Pleia-des, which apparently rules out that explanation!

So on that rather negative note, we must leave the saga of the lost Pleiade. Will a satisfactory answer be found someday, or is this forever destined to be a mystery without solution? If anyone reading this has a bright idea, please don't keep it hidden under a bushel!

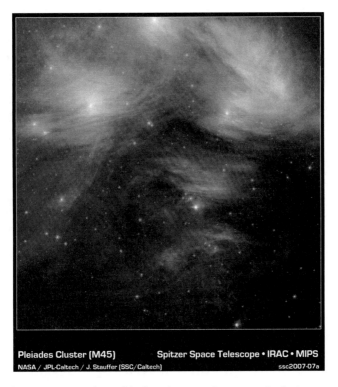

The Pleiades in IR. In infrared light, the enveloping nebula is spectacularly revealed in this Spitza image. NASA/JPL/J. Stauffer (Spitzer Science Center Caltech) NASA/courtesy nasaimages.org.

The Red Sirius Mystery

Sirius, the so-called Dog Star or Alpha Canis Majoris, is the most brilliant true star in the night sky and shines so brightly that it even casts weak shadows under favorable conditions. Other than the extremely rare supernova, the only brighter objects visible at night are not genuine stars. Venus, Jupiter, and (occasionally) Mars are the only planets that, together with the Moon, outshine this queen of nocturnal stars.

The star is actually a close double, although the secondary (Sirius B) is too faint to be seen with the naked eye (and would still be too faint even if removed from the glare of its partner). It contributes only negligibly to the combined light of the pair. The main star is officially known as "Sirius A," and, unless otherwise stated, "Sirius" will refer to this principal member of the duet.

Sirius the Dog Star and its diminutive companion, suitable nicknamed "The Pup". Niko Lang, June 2006.

Sirius is a relatively bright star in its own right, with a luminosity 25 times greater than that of our Sun. At a distance of just 8.6 light years, it is also one of the nearest to our Solar System. It is the combined effect of its intrinsic brightness and proximity to Earth that makes it appear so bright in our skies.

However, Sirius is still a firefly in comparison with the next brightest star in the night sky, Canopus, which has an intrinsic luminosity some 80,000 times greater than the Sun's. Yet, at a distance of 650 light years, this giant is relatively remote. If the positions were reversed, Sirius would be invisible without a pair of binoculars, while Canopus would be about as bright as the quarter Moon – bright enough to blot out the surrounding stars and seriously compromise the darkness of our nights.

Sirius is classified as a white main sequence star of spectra type AIV "Sirius B," colloquially known as "The Pup," by contrast, is a white dwarf, the collapsed and ultra-dense remnant of a once brilliant star.

Sirius A and B. The great difference in the brightness of these stars is readily apparent in this image. NASA/STScI.

The Sirius system is between 200 and 300 million years old – youthful in terms of star ages. (Compare its age with over four and a half *thousand* million years for that of the Sun!)

What is now the fainter Sirius B was once the larger and brighter of the pair. It was a star that lived fast and died young – "died" at least as a hydrogen burning main sequence object. An estimated 120 million years ago, Sirius B shed its outer layers and collapsed into the hot stellar corpse that we see today. During the "shedding" phase, Sirius B would have been a red giant of such brilliance that its light overwhelmed that of its smaller companion, and Sirius (the combined double star) would have shone with a reddish hue.

This, let us repeat, all happened a very long time ago, about 120 million years in the past. There should therefore be no records of a red Sirius. For one thing, the system was a lot more remote at the time Sirius B went from being top dog to being a pup. For another, dinosaurs did not keep astronomical records!

So what are we to make of several ancient writers describing Sirius as red?

Principally, the astronomer Ptolemy in AD 150 named six stars as being red. These stars were Betelgeuse, Antares, Aldebaran, Arcturus, Pollux, and Sirius. The first five do indeed shine with a reddish or orange hue, but Sirius? Surely it is white, clearly not red! This discrepancy was noted as long ago as 1760 by amateur astronomer Thomas Barker, who spoke to the Royal Society in London about the problem.

To make matters worse, the poet Aratus also referred to Sirius as being red in his poem *Phaenomena*. This description was repeated by both Cicero and Germanicus, without comment, in their translation of this poem. Moreover, the Roman philosopher Seneca went so far as to describe Sirius as being of a deeper shade of red than Mars! There is also a possibility that St. Gregory of Tours in the eighth century referred to Sirius as *rubeola* or "reddish," although some scholars think that the star in question may really have been the indisputably reddish Arcturus.

Is it possible that astrophysicists got it wrong about the time when Sirius B left main sequence stardom and became a white dwarf? Is it possible that this actually happened, not 120 million years ago but around the beginning of the first millennium of the Christian era?

All the experts speak with a single voice on this issue.

"Not a chance!"

The astrophysics just does not add up. The time scale is simply too short to accommodate such a change.

Then, as if to deepen the mystery still further, not all ancient writers describe Sirius as being red. The first century poet Marcus Manilius, for example, described it as "sea-blue," and three centuries later Avienus gave it a similar description.

Moreover, the ancient Chinese used it as the standard for white stars, and there are numerous other records dating from the second century BC until the seventh of our era that describe Sirius as being of a whitish hue. Clearly, the "red" records are in the minority during that period.

So if Sirius did *not* look red in ancient times, why did a minority of writers describe it as such? No truly satisfactory answer exists, although a number of suggestions have been proposed.

Maybe, according to one hypothesis, its description as "red" does not actually refer to the color, but is to be taken in a more metaphorical sense. Perhaps it was meant to imply a sign of ill fortune.

This has some support, at least among the ancient Greeks, in so far as the appearance of the star was regarded by them as heralding the hot and dry summer months. Viewed through the unstable atmosphere of early summer, brilliant Sirius twinkled more violently than at other times, and the Greeks believed that this signaled malign emanations that caused people to become "star struck." They even referred to these scintillations as "burning" or "flaming," and it is no great step to characterize these – more or less symbolically – with the color of fire.

It has also been plausibly suggested that the dramatic scintillations of this star, when observed low over the horizon, could genuinely give the impression of flashes of a reddish color. Actually, the scintillations of Sirius can be spectacular as the brilliant point of light goes through the spectrum of colors. When violently scintillating, as at the beginning of a Greek summer, the colors (including red) flash more noticeably and could explain why the name "flaming" was given to it. It may also explain the attention paid to the red flashes.

Several years ago on a clear and dark night this author was amazed to see a brilliant ruby-red star near the eastern horizon. A second later I realized, of course, that it was none other than Sirius, reddened thanks to its low altitude and vigorous scintillation. I recall wondering if this was indeed the controversial red Sirius phenomenon. It may not be the last word on the issue, but seeing the star flash glorious red that night looked convincing to me!

Jumping Jupiter (or Maybe, Jumping Vega?)

The following incident is no joke. It really happened to a colleague of the writer.

It was three o'clock one morning, and my friend was tucked up in bed sound asleep. Then the telephone rang. Still half asleep, our hero managed a sleepy "hello" and was answered (unbelievably) by "Oh, I hope I didn't get you out of bed"!

Now, just how does one answer that sort of greeting at 3 a.m.? Being a quick witted fellow not adverse to a bit of irony, he replied "No, I'm always up and around three in the morning"!

The rest of the conversation went something like this:

There is a light jumping up and down on the northern horizon!
It's a bright star. (looking out the window).
But ... but stars don't jump up and down!
No. But eyeballs do.

(Telephone hung up with a loud "clunk" in the caller's ear! This was before the advent of mobile phones with their gentler disconnections.)

The early morning caller had experienced a phenomenon that can be quite striking – and even a little unnerving – to one who witnesses it. Stars sometimes appear to move erratically while being watched. Bright stars seen near the horizon can appear to jump up and down like a rapid yo-yo, and the effect does indeed look very real. Being near the horizon, it might be thought that some trick of the atmosphere is the culprit, but actually the effect lies wholly with the observer.

The great polymath, Alexander von Humboldt, was the first person to notice – or at least, to take note of – this apparent movement of stars. On a mountaintop just before daybreak one morning in 1799, von Humboldt noted that some of the stars appeared to be performing oscillatory movements that he called *Sternschwanken*. The illusory nature of this was demonstrated many years later by Schweizer, who showed that the apparent movements differed from observer to observer, something that would not happen if they were the result of atmospheric disturbances. Then, in 1887, the phenomenon was given its enduring name; *autokinesis*.

The phenomenon is a well known one among psychologists (the writer recalls being used as a guinea pig for an autokinesis experiment by a friend studying psychology at a university), but there is as yet no agreement as to its precise cause. One popular theory holds that eye movements are responsible (hence my colleague's retort that eyeballs bounce up and down!). According to this explanation, when there are no visual references close to the observed light, eye movements fool the brain into thinking that the light itself is moving.

The problem with this, however, is that several researchers have shown that autokinesis can occur when no eye movements are recorded. Researcher Richard Gregory suggests that when there is a lack of peripheral information, any correcting movement of the eye due to muscle fatigue is wrongly interpreted by the brain as movement of the light.

The apparent movement of a star or planet as seen by an observer on the ground can be startling and may even trigger a UFO report, but when the pilot of an aircraft thinks that a star up ahead, or a known fixed ground beacon, is jumping or drifting, he can very easily misinterpret this as a movement of his airplane. By "correcting" for a non-existent movement, a pilot could inadvertently fly into serious trouble – or even into a mountain! Before pilots were made aware of the phenomenon, autokinesis was blamed for numerous air disasters. Fortunately, with increased recognition of this phenomenon, it has ceased to present the danger that it once did.

Not every movement of astronomical bodies is due to autokinesis, of course. Something even weirder than jumping stars get reported from time to time, and it is to these reports that we now turn.

6. Moving Mysteries and Wandering Stars

Everything in the universe is, one way or another, in a constant state of motion, some things more obviously so than others. Thus, while we would need to wait many lifetimes to see the familiar patterns of the constellations alter as their constituent stars followed their stately cosmic dance, the motions of nearby objects such as the Moon and planets can easily be monitored over a number of days or weeks. Comets and asteroids likewise glide in front of distant stars at a comparatively swift pace, and meteors simply flash past our eyes.

There is nothing weird in any of this, of course. But what happens when we spy something moving through the skies in a way that simply doesn't fit with the usual pattern? What happens when a "star" decides to drift through the constellations at a pace more suited to the average comet? Or something that looks like a comet races across the sky at a rate somewhere between that of the Moon and a slow meteor?

Now that would be weird! But these things surely don't happen – or do they?

As a matter of fact … they do!

Henry Harrison's Puzzle

On the night of April 13, 1879, Henry Harrison of New York was observing the heavens when he noted a most curious phenomenon. He spied what looked like a small comet, except that it was moving across the sky at a rate of over 2 min in right ascension for every minute of time – easily fast enough to watch it move in real time, as seen through the eyepiece of a telescope!

D.A.J. Seargent, *Weird Astronomy*, Astronomers' Universe,
DOI 10.1007/978-1-4419-6424-3_6, © Springer Science+Business Media, LLC 2011

Harrison, surprisingly, appears to have been quite unfazed by his sighting, dismissing the object as a "phenomenon ... of a meteoric nature" and would have been content to leave it as a private curio were it not for a friend whom he called into his observatory to share the experience. This friend, apparently, prevailed upon Harrison to write a short letter to *The New York Tribune* as well as persuading him to contact Professor Hall at the Washington Observatory.

As it turned out, Harrison's approach to the observatory was not received as well as it might have been. Unfortunately, the astronomers noted that Harrison's object was, soon after he found it, quite close to a known comet (Brorsen's) and, apparently overlooking its reported rapid motion, concluded that Harrison had simply "discovered" the latter object. Nevertheless, Harrison specifically noted that he saw *both* Brorsen's Comet *and* the mystery object. Indeed, he watched the object sail past the comet soon after his initial sighting and noted that it trekked over more sky in a few minutes than Brorsen's Comet covered in an entire day!

Harrison watched the object for some 6 h (which actually makes his claim that this was a "meteoric phenomenon" hard to accept), during which time it moved from the region of Andromeda/Triangulum/Perseus to that of Corona Borealis, crossing some 13 h of right ascension!

Apparently, Harrison was not the only person to see the strange intruder. His letter to the newspaper was followed a few days later by a second communication, this time from one Spencer Devoe of Manhattanville, who claimed that he had also sighted the rapidly moving interloper about the same time as Harrison.

What are we to say about this?

Probably the easiest way of handling reports such as these is simply to say that the person making the claim must have been mistaken. This is, in effect, just what the Washington Observatory did, much to Harrington's chagrin!

Nevertheless, as we shall see soon, this report does not stand alone. Several objects of apparently similar appearance to that noted by Messrs Harrison and Devoe have been reported over the years by highly credible observers. It is unlikely that they were all mistakes!

The object's description (scant though it may be) reads like a very tiny comet passing exceedingly close to Earth. It must have come close – *really* close! – to have moved so rapidly across the sky, and the fact that it was not obvious to the world's population as a huge fuzzy mass means that it must have been incredibly small for a comet; but more of this shortly.

It would be helpful if an orbit could be calculated from Harrison's observations, but that exercise is not as straightforward as it sounds. The problem is, the object must have been *so* close that even the large arc of sky over which Harrison followed it still represented but a tiny section of its orbit. Trying to determine its true orbit from this tiny section is a labor fit for Hercules.

Nevertheless, an attempt was made in the early 1980s by Charles Townsend and Scott Hanssen, and, although no actual orbit was forthcoming, their work indicated that a short-period ellipse was more likely than a nearly parabolic orbit. They also confirmed that it must have passed very, very close to Earth.

At the author's request, another attempt was made soon thereafter by David Herald of Canberra, Australia, with some interesting results. After some fiddling with a computer program that he had written several years earlier, he managed to get something that looked like a realistic comet orbit from Harrison's positions, although he stressed in the strongest possible terms that it should not be taken as other than very approximate at best.

So with this caveat ringing in our ears, what can be said about the Herald orbit?

Well, it actually looks very like that of a short-period comet, in agreement with the Townsend-Hanssen result. Because Harrison's positions were only approximate and because of the problem mentioned above, Herald had to assume a parabolic orbit, but the results he got indicated something of very low inclination to the plane of the ecliptic. This is very typical of short-period comet orbits. The computed figure of 3.2° inclination should not be pressed too far, but it does suggest that the angle was very small.

The Herald results also suggested that the object was closest to the Sun around March 19 at a distance of 0.9 times that of the radius of Earth's orbit. Once again, these figures need to be taken with a large grain of salt, but they do suggest that it was moving away from the Sun when Harrison saw it.

The orbit computed by Herald further indicated an extremely close approach to Earth (not surprisingly!) about the time of

Harrison's sighting. In fact, it indicated that the object passed just 37,000–38,000 miles (around 60,000 km, give or take a few) from Earth. Although this figure should not be made to carry too much weight, it is probably not too far from the truth. Actually, of the orbital elements calculated by Herald, the one determining the object's crossing of Earth's orbital plane – and therefore, in these circumstances, the closest approach to our planet itself – is the best determined. We can probably take this figure as a ballpark value at least.

Unfortunately, except for the allusion to the object as "comet-like" (i.e., nebulous), Harrison provides little by way of physical description. He says nothing of its size or brightness, although the fact that he mentions inviting his friend into his observatory to look at it suggests that it was only visible through a telescope. Moreover, had it been clearly visible with naked eye, we can presume that many more people would have noticed it.

Moreover, it may (or may not!) be significant that Harrison did not criticize Washington Observatory's bungled "identification" of the object as Brorsen's Comet by insisting that what he saw looked very different from this comet. This cannot be taken too far, of course, but it might be supposed that if it *did* look very different from the comet, he would have mentioned that fact as well. At that time, Brorsen's Comet appeared as a fuzzy ball about one-sixth of the Moon's diameter and was just a little too faint to be seen by eye. Maybe Harrison's object was not too dissimilar in appearance.

If its apparent size and brightness did roughly match that of Brorsen's Comet, and if its distance from Earth was close to the value suggested by the Herald orbit, the real diameter of this nebulosity would have been only some 56 miles (90 km), and its true or intrinsic brightness around 240 *million* times fainter than Halley's Comet during its 1986 return!

Whereas normal comets possess nuclei of several miles diameter, surrounded by gaseous "heads" ranging from tens of thousands to hundreds of thousands of miles across, the entire "head" of Harrison's object may not have been much larger than the nuclei of some of the larger comets. If the nucleus/head ratio held, the solid core of this comet (if that is truly what it was) would seem to have been measured in yards, if not in feet or even inches!

Something as small as this and composed mostly of ice could not survive for long within the inner Solar System, and some people have seen this as an objection to the "comet" interpretation of Harrison's sighting. Larger icy fragments are not infrequently shed by comets passing near, or within, Earth's orbit, and these fragments generally fade out in days if not in hours.

On the other hand, since it began its solar monitoring back in 1996, the SOHO satellite has beamed back images of large numbers of very faint comets making dangerously close approaches to the Sun. Many of these have been estimated as clocking in at less than 10 yards diameter. Few seem to be as large as 100 yards.

Admittedly, the smaller ones and those passing less than about a million miles from the Sun's surface boil away and disappear, but some that do not pass quite so close (though still well within the orbit of Mercury) not only survive but have even been found to be following periodic orbits that bring them back into this roasting environment every few years. Some are found to have periods of barely four years. These comets must be very durable, and it is not impossible that Harrison's object was one such tough little critter – perhaps only a few tens of feet across, maybe even less – that remained even better preserved, thanks to its greater minimum distance from the Sun.

It is also worth noting that calculations by Martin Beech and Simona Nikolova modeling the durability of blocks of ice within the inner Solar System indicate that an ice ball of 10 m (about 11 yards) diameter and moving in the orbit of Comet Tempel–Tuttle, the parent of the Leonid meteors, would take about 1,900 years to completely evaporate away. In the orbit of Tempel–Tuttle, this hypothetical ice ball would approach the Sun to about the same distance as Herald's orbit suggests for the Harrison nebulosity, although the period of 33 years is probably considerably longer than that of the latter. Nevertheless, the Beech–Nikolova model also suggests that an ice ball of similar size and moving in the orbit of Comet Encke – approaching the Sun to within Mercury's orbit every three years – would still manage to endure for half a century or thereabouts.

Taken together, these considerations make the Harrison report seem a little less weird and a lot more probable.

Franz's Fuzzy

As remarked earlier, the strange case of Henry Harrison does not stand alone.

On July 5, 1911, J. Franz, an astronomer at Breslau, was searching for Kiess' Comet when he came across a nebulous blob moving across the sky at the rate of 3 min in right ascension for every 6 min of time. Initially, Franz thought that he had found the comet he was seeking, but the rapid motion soon put paid to that idea. Moreover, he did find Kiess's Comet shortly afterwards, eliminating all possibility of misidentification.

Unfortunately, Franz did not follow the mystery object for long, probably because his attention was drawn to the real subject of the night's observing – Kiess's Comet. Yet he did give a description, noting that it was about 6 min of arc in diameter and of sixth magnitude, that is to say, about one fifth of the Moon's apparent diameter and barely on, or slightly under, the naked-eye limit. He also noted that the distance of the mystery body (assuming an essentially parabolic orbit) could be found by means of a simple formula relating the angle made by its motion to our line of sight and the distance of the Moon. Depending on the value of this unknown angle, the object was, either, almost on our rooftops (for motion very close to the line of sight) or about 625 thousand miles (one million km) away if it moved perpendicular thereto.

Taking the latter value, his description suggests that the nebulosity was nearly 1,250 miles (2,000 km) in diameter and about 400,000 times fainter, intrinsically, than Halley's Comet. As its motion was probably not exactly perpendicular to the line of sight ,and the orbit was likely an ellipse rather than a parabola, there is a good chance that its real distance was less than this absolute maximum value, and its size and brightness correspondingly smaller.

Both this object and the earlier one seen by Harrison are, most likely, best explained as tiny comets passing very close to our planet. The following case seems at first glance to suggest a similar explanation, but in this instance all may not be as it seems …

Wilk's Fast-Moving Mystery

On September 1, 1926, a cablegram from Professor E. Stromgren announced the discovery of something very strange – an apparent comet moving through the skies at the rate of 1° every 4 min!

The discovery was made, according to the message, by A. Wilk at Cracow in Poland, and the object was described as being "oblong" and of sixth magnitude, just at the limit of naked-eye visibility under favorable conditions. Wilk was using a pair of 7×50 binoculars as part of his regular scan of the night sky when he first sighted the nebulosity, but he also located it in 3-in. (8-cm) comet-seekers at powers of 12, 20, and 40 times.

Once again, this was an instance of a seemingly incredible discovery made by an entirely credible discoverer. Wilk was no stranger to comets. In fact, he had discovered one the previous year and was to have his name given to three more before the end of the following decade.

At least one astronomer, on receiving the notification, wondered if the reported motion (equivalent to 15 degrees per hour!) was actually a misprint for 15° per day (still unusually fast), but the reported movement turned out to be real and implied a distance of less than twice that of the Moon. If, indeed, the sighting was truly of an astronomical body.

Unfortunately, no other observations of the fast-moving object were made, and some astronomers – while in no way disparaging either Wilk's integrity or the accuracy of his report – began to wonder whether what he had seen was astronomical at all.

The reason for their skepticism, apart from the sheer speed at which the thing was moving, was the close match between its velocity and direction and that expected for a stationary object in Earth's atmosphere, projected against the daily motion of the sky. Maybe what Wilk saw was really something in our own atmosphere projected against the background stars. A. C. D. Crommelin suggested that it might have been a persistent meteor trail (seen more or less end-on?), but George van Biesbroeck of Yerkes Observatory noted that, as seen from Wilk's position at Cracow, the object would have been just 4° above the southern horizon. Might it have been a terrestrial light source such as "a pilot light on a captive balloon, automobile light on a mountain, [or] mirage of a terrestrial light" that looked somewhat comet-like in Wilk's instruments?

Of the various possible explanations suggested, this present writer tends to favor that of Crommelin, that is to say, a distant and persistent meteor trail. The case for a close encounter with a small comet, though certainly not excluded, is much weaker than the analysis of Harrison and Franz.

These above three reports, although strange, are still not altogether beyond explanation in relatively conservative terms. Two very small comets and an odd meteor trail is what we suggest as the most likely set of culprits, while leaving the door ajar for other possible explanations.

The next two instances of "comet-like" sightings are, however, more difficult to explain either as real (if atypical) comets or as meteoric phenomena. Not that either explanation is totally excluded. They just seem a little forced!

PROJECT 19
If You Should Spy a Moving Fuzzy...

If you spend a good deal of time under the night sky, there is a chance that sooner or later you will see something fuzzy moving through the heavens too quickly for a "regular" comet, yet too slowly for a nebulous meteor. Chances are the object you see will be a satellite being maneuvered into a slightly different orbit.

Yet, the experiences of Messrs. Harrison, Franz and (maybe) Wilk – all of whom lived before the advent of artificial satellites – argues that (very occasionally) a mystery fuzzy might be something natural. So if you should see something, confirmation from another observer would be beneficial, not just to corroborate your story, but (if good measurements are taken) to derive at least the approximate distance of the object. Photographs or video footage would also be good, especially if these could be obtained from different locations.

If you own a Swan Band filter, check to see whether the object is easier to see when viewed through this. These filters transmit the typical cometary gaseous emissions, and those comets whose light mostly comes from gas (i.e., those having a low dust content) appear brighter when viewed through such a filter.

If your nebulosity is enhanced, it might really be a tiny comet passing very close to Earth. On the other hand, if it remains unchanged or appears fainter, a cometary nature is not *automatically* ruled out. It may simply be a dusty comet where sunlight scattered off solid particles overwhelms the gaseous emissions. Something artificial is, however, far more likely.

Unless you are thoroughly convinced that what you are seeing is not an artifact of the Space Age, it is best not to be too public about your observation until you have checked out all known satellites in that region of sky. Even then, secret military or spy satellites are hard to rule out, other things being equal. Report it certainly, but remain conservative in your assessment of what you report it as being.

A Bright Streak in the Cordoba Sky!

On the morning of May 5, 1916, A. E. Glancy and C. D. Perrine of the National Observatory at Cordoba, Argentina, discovered a most unusual phenomenon. It appeared, in the words of Perrine, as "a bright streak just below Alpha Pavonis, sensibly straight, about 8° or 10° long and one-half to one degree in width. It was more sharply defined toward the west, that extremity resembling the head of a large bright comet, but without any well defined condensation or nucleus ... it was an exact counterpart on a smaller scale of Halley's comet when this object was near to the Earth [in 1910]."

Both astronomers kept the object in view for just over 1 h, during which time it had moved some 10° in the direction of the Sun. They observed it with binoculars, the finder of the observatory's 12-in. (30-cm) refractor, and with the naked eye.

Perrine noted that it grew fainter as it neared the horizon and that "the diurnal motion [was] not sufficiently counteracting its own motion."

He also commented that the "tail" was initially "without steamers" but that one began to form around the middle of the period of visibility, presumably about half an hour after Glancy's first sighting, and remained visible until the object went out of sight.

What could this strange apparition have been?

Perrine suggested that it was probably either an unusually persistent meteor trail or a comet passing very near Earth on its way toward the Sun. He favored the second alternative, reasoning that a meteor trail should have contorted significantly during the period of his and Glancy's observations. Moreover, one may add that the only change that *was* noted – that of a streamer forming – is hardly what one would expect to see in a meteor trail.

Perrine certainly should have known a comet when he saw one. He had been observing these bodies for many years and was officially accredited with the discovery of nine. His comment must, therefore, be taken seriously.

Yet, something is not quite right. Clearly, the object (whatever it might have been) was very close to Earth. Nevertheless, it did not really *look* like a comet seen at very close quarters. Most of these objects display large and diffuse heads at close range, but this thing appeared to be mostly tail. That is not a knock-down drag-out objection, of course (some comets have atypical appearances), but it does make us wonder ...

Nevertheless, on the assumption that it was a comet passing very near to Earth, Glancy calculated an orbit from three positions that she and Perrine had taken, warning, however, that "Too much confidence should not be placed in such an orbit as this." The orbit turned out to be elliptical with a period of just over 16 years. Not surprisingly though, nothing was seen in 1932!

If the mystery object was not a comet nor a meteor trail, is there anything else that it might have been?

We might recall that, during the earlier discussion of peculiar meteors, mention was made concerning the observation of dust swarms by artificial satellites monitoring the dust environment of near-Earth space. These swarms, it was said, apparently resulted from the electrostatic disruption of large but very fragile meteoroids passing through the auroral zones of our planet. Now, it may appear far-fetched (and may be as far-fetched as it appears!), but perhaps a large meteoroid in the process of disruption beyond Earth's atmosphere left a train of fine dust behind it; fine enough to be swept back by the pressure of Sunlight, in effect imitating a comet's tail.

Such fine dust is a good reflector of sunlight and, perhaps even more importantly, a good forward scatterer of sunlight if the geometry is right, i.e., if the dust is more or less between the observer and the Sun. This forward-scattering effect is what makes windborne thistle down and spider webs suddenly visible as they pass in near line with the Sun, as mentioned in Chap. 2 of this book. Several studies show that this effect is capable of causing tremendous enhancement in the brightness of comets seen in the immediate vicinity of the Sun, but on the earthward side of it.

Is it possible that what Glancy and Perrine saw was a large but extremely fragile dust-bunny of a meteoroid in the process of being blown apart by electrostatic changes resulting from its passage through Earth's radiation belts?

The reader is left to ponder that one!

Edie's Enigma

Lest we might think that the above entry is about as weird as it gets, we might like to try figuring out the nature of the phenomenon observed by Cape Town astronomer L. A. Edie on October 27, 1890.

At 7.45 p.m. local time, this astronomer saw what he described as a comet sporting a naked-eye tail of some 30° in length almost due west of his location. The tail was described as tilting to the south at an angle of about 45°.

Movement was detected almost immediately with the naked eye, something rare for comets. The phenomenon moved around the western and southern horizons at an altitude of between 20 and 25° as the "tail" grew longer, eventually reaching some 90° (half the span of the heavens), yet remained just 1° in width and running parallel with the southern horizon. It remained in view for 47 min before fading into the southwest.

It has been suggested that this object – almost certainly *not* a comet – was auroral in nature. That explanation is not watertight, but unless a reader can think of a better one, it will have to do for now!!

John Dove's Mobile Mystery

The next account is interesting both for being the oldest report given here as well as for the fact that it was first reported to none other than Edmond Halley and published as a letter in the 1732 September-October issue of the *Philosophical Transactions of the Royal Society of London*. It therefore *should* come with an excellent pedigree!

The observer was John Dove, who at the time of the observation was on board the *Monmouth* in the southern Atlantic, just west of Africa. The observation was made on February 29, 1732.

Dove writes that "the Moon shining very bright, being near the Full, we saw something very bright rise about West, which I judge to be a Comet: It set about East, passing from West to East in about 5 min, between the Moon and our Zenith, and to the Southward of Spica Virginis, it carried a Stream of Light after it about 40° long and 1° or [1.5°] broad; the Brightness of the Moon outshined the Comet as it came near it."

The most likely explanation for this sighting is a very bright and unusually slow fireball. We can be pretty sure that it was not a comet, and it seems that Dove made the not-too-uncommon mistake of confusing bright meteors (complete with long tails) and comets. Admittedly, the time span of 5 min does seem unusually long for even a very slow fireball, but the splendor of the event and its uncommon appearance may have so distracted the witnesses that a true sense of duration was lost.

By the way, the remark that the Moon "was near Full" is not confirmed. A full Moon did not occur until March 11, so either the date is wrong or the Moon was not even first quarter. This alone should warn us against placing too much heed on the "five minutes" of visibility.

Coincidentally, the 1813 issue of the *Monatliche Correspondenz* relates an alleged sighting of a comet above Spica on February 27, 1732, by a Mr. Hanow (possibly in Danzig). No other details are forthcoming concerning this, but it seems very unlikely to have been related in any way to the Dove object.

The Wandering Star of Hofrath Huth

Comet-like objects are not the only mobile mysteries to have been reported at various times. There was also the strange incident of the "moving star" observed by Hofrath Huth in December, 1801.

In a letter to Bode dated December 5, Huth wrote "In the night from second to the third of this month, I saw with my 2-1/2-ft Dollard, in a triangle ... to the southwest, a star with a faint reddish light, round, and admitting of being magnified. I could not discern any trace of it with the naked eye; it had three small stars in its neighborhood." He wrote again 10 days later saying that, because of unfavorable weather, he had only managed three further sightings of the "star" – during the early morning hours of December 3, 13, and 14, and that he had deduced that the object had a retrograde motion toward the southwest. From his measurement on December 14, he estimated that it had moved 4 min of arc since the previous morning and no more than 30 min (about the breadth of the Moon) since December 3.

A further letter written on December 21 mentioned that the "star" had been seen on only one more occasion – the night of December 19/20 – when Huth found it "near four stars apparently situate to the westward, about half a diameter of the full Moon below the smaller one." It appeared to be moving toward i Leonis and toward the ecliptic. Huth continued: "Of the motion of this planet-like star I can no longer doubt, since I have observed a difference of 5/6° nearly, between its positions on the 3rd and 20th."

Bode received a fourth letter from Huth on January 12, 1802, mentioning two further sightings of the object on January 1 and 2. In that communication, Huth mentioned that the "star" appeared even smaller (presumably meaning fainter) than one of the satellites of Jupiter, and on the second night he had difficulty seeing it due to its proximity to another star. On January 5 he could discern it only occasionally, and on the following night, there was no trace of it.

So what was Huth's moving "star"?

Apparently, it was not a comet, as at no time did Huth mention any misty head or coma surrounding it. Even a very small and condensed comet would almost certainly have betrayed its nature by showing just a little fuzziness!

A superficially more probable suggestion is an asteroid. We must remember that in January 1802, only one asteroid or minor planet was known. This object, Ceres, was found the previous year by Giuseppe Piazzi. Confirmed by other observers in early 1802 it, unlike Huth's object, was quickly added to the Sun's known family. The second known asteroid, Pallas, was discovered on March 28, 1802, and the orbits of the two calculated by German mathematician Karl Gauss (who, by the way, referred to them rather disparagingly as "a couple of clods of dirt which we call planets").

Nevertheless, identifying Huth's object with an asteroid raises its own set of problems.

For one thing, *which* asteroid could it have been? True, the number of asteroids is huge, and new ones continue to be discovered at an amazing rate, but with very few exceptions, these are extremely faint bodies that could not possibly be the object seen by Huth. Although he mentioned that his "star" was not seen with the naked eye, it could not have been much fainter than a dim naked-eye star. As we have seen, he noted that it was smaller (or fainter) than one of the satellites of Jupiter in early January. The four "Galilean" satellites of this planet are each technically bright enough to be seen with the naked eye (as we shall see in the following chapter of this book), and the brightest would be widely observed without optical aid, were it not for the close proximity of brilliant Jupiter. To judge something as fainter than one of these moons still hints at its brightness being in the same approximate ballpark. One would hardly mention that X was fainter than Y if their respective brightnesses differed by a factor of a thousand, for instance!

Moreover, Huth apparently detected some color in the object. To stimulate color vision, it must have appeared pretty bright in his telescope.

From this, we can conclude that the object was not very much fainter than a marginal naked-eye star, which actually makes it at least as bright as Ceres and probably in the same region as the brightest asteroid, Vesta. Oddly, although Ceres is the largest of the asteroids, it is not the brightest. That honor belongs to the smaller but more reflective Vesta, the only "main belt" asteroid that can, on very favorable occasions, be seen – albeit faintly – without optical assistance.

After Vesta and the other "bright" asteroids were discovered, it was easy enough to trace their paths backward through time and see if they were near the position of Huth's star in 1802.

They were not!

There is another problem. Huth mentioned that his object "[admitted] of being magnified." Not even Ceres and Vesta look other than point-like in ordinary telescopes. In fact, the term "asteroid" just means "star-like" and was given to these bodies because of their appearance even under high magnification. In their true nature, they are anything but starlike!

If Huth's object really did appear larger under higher magnification, it must have been very large and quite close – a size and distance combination unprecedented in asteroid observation. We now know about a population of near-Earth asteroids that, in theory, could become naked-eye if large enough and close enough. Something that close would be moving a lot faster than Huth's description implied, however. If it was near enough to show a disc at high magnification, it was *too* close!

There is another possibility, although it is a bit farfetched.

In recent years, several long-period asteroids-like bodies have been found, as well as some large icy objects, at the edge of the Solar System. These may be dormant comets rather than true asteroids, and some have indeed revealed weak cometary activity as they drew nearer the Sun. Is it possible that one of these dormant objects swept through the Solar System in 1802 and was picked up by Huth. Interestingly, these bodies also tend to be reddish in color, and if one became bright enough, it could appear slightly reddish in a telescope – just as Huth described.

Of course, there is another explanation; one which has gained quite a following in fact.

Huth made the whole thing up!

This may be true, but it would seem a pretty ill-advised thing for him to do. What could he gain by such a fabrication? And why pretend to find something as controversial as a moving star, especially at a time when only one asteroid was known? Of course, a cynic might say that he coveted the chance of being the second asteroid discoverer or even that he secretly wished to mimic Herschel's feat of finding a major planet. Yet feigning discovery of a planet or an asteroid would harm rather than benefit the reputation of someone like Huth. His "discovery" would never be confirmed, of course, and his name would forever remain under a cloud. Why would anyone wish that on himself? It seems more reasonable to conclude that the object was real, even if it is not very readily explained.

Bullseye for 2008 TC_3!

This moving object is different from those featured in the above stories. There is no question about what it was. The tale of its discovery and subsequent events is unusual for a different reason, one which will soon become obvious!

On October 6, 2008, R. A. Kowalski of the Catalina Sky Survey at Mt. Lemmon discovered a faint fast-moving asteroidal object. Nothing unusual about that, of course. After all, the survey is *supposed* to find asteroids that might someday pose a danger to Planet Earth.

But this asteroid was close. *Very* close. Indeed, when discovered it was just 1.27 times the distance of the Moon – and still approaching!

An orbit was quickly calculated, and this showed that the newly discovered body (now designated 2008 TC_3) was heading straight for us and would hit the Earth on October 7 at around 2 h 45 min Greenwich Mean Time. It would enter our atmosphere over the Sudan while moving in a west/east direction.

Fortunately, the announcement caused no panic. Even more fortunately, there was no need for any. The asteroid was tiny, only about 15 ft (4.6 m) in diameter and weighing in at an estimated 80 tons or thereabouts. It was expected to simply burn up in Earth's atmosphere, with nothing more than dust reaching ground level.

Following the discovery of 2008 TC_3, astronomers watched the asteroid's approach as it rapidly brightened. By early October 7, it was probably bright enough to find in large binoculars if one new exactly where to look, but about 1 h before the time of predicted impact, it vanished from sight. No, nobody shot it down; it simply entered our planet's shadow and remained in eclipse for the final leg of its doomed journey.

Because it encountered Earth's atmosphere over a sparsely inhabited region, few reports of its fiery plunge are known. However, J. Borovicka of the Astronomical Institute, Czech Academy of Sciences, reported that Z. Charvat noted a bright spot on images taken by the weather satellite *Meteosat 8* at the predicted time of the encounter. This corresponded, as near as could be determined, with the expected fireball. The object was trailing what appeared to be a tail of about 1.8 miles (3 km) long toward the west/northwest,

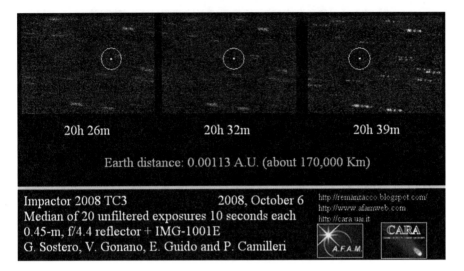

Doomed asteroid 2008 TC_3 approaches Earth as captured on these three images. Courtesy G. Sostero, V. Gonano, E. Guido and P. Camilleri.

consistent with something hurtling through the atmosphere from an approximately westerly direction.

Moreover, sensitive barographs dedicated to monitoring large atmospheric explosions picked up the shock from an aerial detonation of between one and two kilotons magnitude in the same place and at the same time as the asteroid's dive into our atmosphere.

Should we call this an instance of an asteroid (albeit – fortunately – a very small one) colliding with Earth or a large meteoroid found well in advance of entering the atmosphere?

Call it whichever you like. It was both. The object (let's use that neutral term for now) blew itself to pieces well above Earth's surface; some 23 miles (37 km) from the ground, according to the best estimates.

Initially, it was thought that no fragments survived to fall as meteorites. First predictions, as already mentioned, thought that the body would be completely consumed within the atmosphere. Nevertheless, this assumption was questioned by Peter Jenniskens of the SETI institute. He turned out to be right, and, in collaboration with Mauwia Shaddad of the University of Khartoum and several students and staff of that university, his team managed to collect a total of nearly 280 fragments strewn across more than

Bright against a dawning sky, the dust trail left as 2008 TC_3 plunged through our atmosphere graces the heavens over the Sudan on October 8, 2008. Courtesy P. Jenniskens (SETI Institute/NASA Ames), M. Mahir and M. Shaddad (University of Khartoum).

18 miles (29 km) of the Nubian Desert during an expedition to the region in December 2008. The total mass recovered was 8.7 lbs (3.95 kg), a far cry from the initial mass of the body, most of which was ablated away into the spectacular dust trail during the asteroid's fiery flight through our atmosphere.

This marks the first time that an object destined to pepper a region of Earth with meteorites has been discovered in outer space, and it seems to give hope that there might come a time when the patrol of the skies reaches a degree where most meteorite-producing bodies are picked up in advance and teams are dispatched to the predicted fall areas prior to the meteorites' arrival. Thus far finding freshly fallen meteorites has always been a hit or miss affair (literally!), but could this be about to change? Advance warning would be especially valuable in the case of the very fragile Type I carbonaceous chondrites, which weather away very quickly but which also contain interesting pre-biotic organic matter. Finding these as soon as they fall would not just save them from weathering but would also greatly reduce the chance of confusing organic contamination from the ground with material indigenous to the meteorite itself.

Maybe someday this will be a reality, but probably not in the near future. As Jenniskens points out, objects the size of 2008 TC_3 strike Earth with a frequency of about one per year, yet this was the first time that one had been picked up in space in advance of entry into our atmosphere. Evidently, the coverage is still far from sufficient to record more than a small percentage of these smaller objects and will likely remain so for some time to come.

In the case of 2008 TC_3, however, the ability to locate the fallen meteorites turned out to be very important. Just before the asteroid entered Earth's shadow and disappeared from view, astronomers at the La Palma Observatory in the Canary Islands determined from the manner in which sunlight reflected off its surface that this was one of the rather mysterious F-class of asteroids.

A lettering system for asteroids was developed during the early 1970s. It is based upon analysis of the spectrum of sunlight reflected from the asteroid's surface. From the details of this spectrum, certain facts can be determined about the mineral composition, at least of the surface, of the body. According to this classification, asteroids are known as S (stony), C (carbonaceous), M (metallic), F (flat and rather featureless spectrum), R (red), and so forth. F types are dark objects with a low albedo (reflectivity) somewhat similar to the C class.

One of the aims of asteroid/meteorite research is to relate the various classifications of meteorites with those of asteroids. Certain progress has been made in this field, but it has not been rapid.

No meteorites had definitively been associated with the F class. But that was just about to change!

The meteorites, which fell in the Nubian Desert, are now officially known as the Almahata Sitta Meteorite (Arabic for "Station 6" – the nearest railway station to the fall) and belong to the rare class known as ureilites. Meteorites of this type are distinguished by being carbon enriched and show clear evidence of having undergone at least a partially molten phase while still within their parent body. They are also known to contain microscopic diamonds, not surprising for carbon-rich objects that have experienced significant heat and pressure.

It is thought that this phase of high temperature and pressure resulted from a collision between the ancestral body of these meteorites and some other asteroid in times long past. During the fairly brief burst of high temperature, oxidized iron particles in the outer parts of olivine crystals were reduced to their metallic form. Some of the indigenous carbon apparently acted like coke in a blast furnace, part of it being converted in the process to CO and CO_2 and escaping into space. Other portions of the carbon (originally in the low-pressure graphite form) were converted into the high-pressure crystalline form of the element – in other words, diamond. As meteorite expert Robert Hutchison expressed it, these meteorites were born in a sort of diamond-forming blast furnace in space.

The finding of the Almahata Sitta Meteorite marked the first time that samples of an F-class asteroid could be examined in the laboratory and solved the mystery of the origin of this rare meteorite class.

Even so, the Almahata Sitta Meteorite is unique among known ureilites in so far as its carbon content has been "cooked" to an unusual degree. It had obviously been exposed to a lot of heating within the parent body. Maybe its parent body – or grandparent or earlier ancestor – experienced an unusually severe impact.

Presumably, the original parent was formed within the Asteroid Belt between Mars and Jupiter and experienced the collision that "cooked" its contents in that location. Over long periods of time, further collisions fragmented the first generation of material into a greater number of ever-smaller pieces. Some of these were eventually deflected into orbits crossing that of Earth, where even smaller pieces were broken away by impacts with meteorites. Asteroid 2008 TC_3 was presumably one of these smaller fragments. The most likely immediate parent is suspected as being the Earth-crossing asteroid 152679, a.k.a 1998 KU_2. This F-type body is some 1.6 miles (2.6 km) in diameter. Fortunately, it was only a tiny chip off this old block that hit us in October 2008!

Undoubtedly, much information remains to be uncovered from the asteroid fragments now in the hands of scientists. But already one important fact has emerged. This tiny asteroid – and probably F-type asteroids in general – was very fragile. This was demonstrated by the relatively high altitude at which it broke up. If one day one of its larger and more menacing siblings should be discovered with Earth's number on it, the action *not* to take would be trying to deflect it with a large explosive such as an atomic bomb. This would only succeed in breaking it into a swarm of smaller, but still dangerous, pieces. Instead of our planet being hit by a single bullet, it would be struck by a blast of shotgun pellets. It is debatable which would be the worse scenario. Some gentle

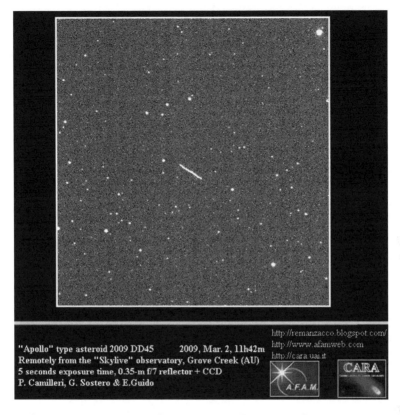

"Apollo" type asteroid 2009 DD45   2009, Mar. 2, 11h42m
Remotely from the "Skylive" observatory, Grove Creek (AU)
5 seconds exposure time, 0.35-m f/7 reflector + CCD
P. Camilleri, G. Sostero & E.Guido

http://remanzacco.blogspot.com/
http://www.afamweb.com
http://cara.uai.it

Asteroid 2009 DD_45 narrowly misses Earth on March 2, 2009. Courtesy P. Camilleri, G. Sostero, E. Guido.

nudging out of the way would be a better choice. But hopefully, this is a decision we will never be forced to make!

As a final thought, it is worth mentioning that just five months after the 2008 TC_3 event, an asteroid measuring about 70 ft (21 m) across blitzed Earth at a distance of just 40,000 miles (64,000 km). This is approximately one fifth of the distance from Earth to Moon. Fortunately, this one missed, as it would have made a lot bigger crash than the diminutive 2008 TC_3 had it actually struck. This object, now known as 2009 DD_45, was discovered by Rob McNaught at Siding Spring Observatory in Australia just three days before the time of nearest approach. At its closest, it was bright enough to be within the reach of a 3-in. (7.5-cm) telescope.

7. Facts, Fallacies, Unusual Observations, and Other Miscellaneous Gleanings

Our ramblings through the byways of astronomical reports, observations, and speculations have brought us to a final collection of odd facts and curios that are worth including in our compendium, yet do not neatly fit into the categories marked out by previous chapters. This final chapter is therefore a mixture of interesting observations and anecdotes of "any other variety," ranging literally from the sublime to the ridiculous and everything in between. From a browse through some accounts of difficult observations, to tales of fossil-like structures in meteorites to apparent blobs of gunk falling from the skies, from the anti-establishment views of Dr. George Waltemath and his crazy moons to that perennially interesting story of the Christmas Star, this chapter will range almost as far and as wide as astronomy itself.

Difficult and Unusual Observations

Storm Observing

Lest you think that we have left astronomy for meteorology, you should know that this section is not about observing storms but about making astronomical observations *during* a storm!

Under normal circumstances, the recording of 47 sunspots in 13 groups would not be included in a list of difficult or unusual observations. But it is not the *subject* of this observation that makes it unusual. Rather, it is because it was made through the eye of a tropical storm!

The observation was by D. W. Rosebrugh from St. Augustine in Florida on June 6, 1968, as the eye of tropical storm Abby (rated as a full hurricane only a day or two earlier) passed across the observer's location.

As the eye of the storm passed over, Rosebrugh had some 20 min of clear sky before cloud and heavy rain once more set in, and in an effort that truly gauges the determination of some amateur astronomers, he used this break in the weather to secure his observation of the Sun!

Moons of Jupiter

The four largest moons of Jupiter, also known as the Galilean moons, in honor of their recognition by this famous astronomer, are reasonably bright objects, and at least one of their number would be relatively easy to spot with the average naked eye if it were not for the overpowering glare of the brilliant planet itself.

Nevertheless, for those with especially keen eyesight and no astigmatism, naked-eye observations of at least the brightest moon should not prove too difficult as long as Jupiter is hidden behind some projection.

Not altogether surprisingly, naked-eye sightings of one or more of the moons have indeed been noted from time to time. Although not officially discovered until Galileo pointed his telescope in Jupiter's direction, it seems that the first recorded sighting of a Jovian moon took place as long ago as 364 BC when Chinese astronomer Gan De noted a small star "in alliance" with the planet. This claim has actually been tested at the Beijing Planetarium, where it was confirmed that people with good eyesight were able to discern star-like objects as faint as magnitude 5.5 just 5 min of arc from something as bright as Jupiter. Three of the Galilean moons – Io, Europa, and the even brighter Ganymede – are brighter than this and should therefore be within the range of people with excellent vision.

Since the moons' discovery by Galileo, there have been numerous alleged naked-eye sightings. Some have been downright fraudulent, others merely possible, and still others highly credible.

In the first category, we have the case of two sisters who lived in Hamburg during the early 1800s and who claimed that they could see the satellites with unassisted vision. The ladies were, regrettably, fakes. All they were doing was reading the positions from drawings given in *Berliner Jahrbuch*!

The satellite configurations that they claimed to be seeing gave the show away. Every one of their reported configurations turned out to be the reverse of what was actually happening. What the sisters failed to realize was that the drawings in *BJ* were for astronomical telescopes, in which images are inverted!

Ironically, the very thing that exposed the sisters' fraud appears to have verified a more recent naked-eye sighting of all four Galilean moons! About 1970, E. Talmadge Mentall of Dorchester, Massachusetts, was standing on the back porch of his home with his daughter Valerie, then about 8 years old. They had just finished putting together a small refracting telescope and were about to try it out on one of the favorite objects for new owners of small telescopes – the planet Jupiter.

Mentall explained to his daughter that when she looked into the eyepiece, she would see four star-like objects extending in a line pointing upwards from the planet. These were Jupiter's moons, he explained. However, while her father was explaining this, Valerie interjected "No Daddy, the moons go *down* from Jupiter!" But she was not looking into the eyepiece (where directions were inverted). She was looking directly at Jupiter in the sky!

Another possible instance of the naked-eye sighting of one of the moons was told by Russian explorer Ferdinand Wrangel of a Siberian hunter who (as he so colorfully phrased it) once saw Jupiter "swallow a small star next to it and vomit it up shortly afterwards"! Whether this was a bona fide sighting of one of Jupiter's moons passing into and out of the planet's glare or whether it was a close appulse of a star is not known. It would be handy to know the date of this event, as the position of Jupiter's moons as well as the position of the planet itself in relation to background stars could then be determined, but without this information, the report remains a possible sighting only.

Several well known astronomers have apparently sighted at least one of the moons by eye alone. The great E. E. Barnard, whose acute vision is legendary, apparently managed the feat, as did his British contemporary W. F. Denning. T. W. Webb claimed that he could detect at least one of the satellites when he was wearing his spectacles, which could more or less pass as a naked-eye sighting. Last century, astronomer Richard Baum also joined the ranks of those who saw at least one of the Jovian moons sans optical aid.

PROJECT 20
Naked-Eye Moon Watching

The biggest difficulty in seeing the brightest of Jupiter's moons without optical aid is the presence of Jupiter itself. If your eyes have any degree of astigmatism, the planet images "flares" into a small blob like a tiny Christmas decoration, and seeing anything next to it is all but impossible. The difference in brightness between Jupiter and its brightest moons is also large, with even Ganymede being close to 700 times fainter than the planet.

The best time to try for a sighting of one or more of the moons is when Jupiter is high, the sky is very clear and dry, and the moon being sought is near maximum elongation from the planet. Ganymede is the "easiest" to find. It is the brightest and, although its distance from Jupiter is significantly less that that of Callisto, it is a full magnitude brighter. Io is estimated as the next brightest (after Ganymede), but remains close to the planet. Ganymede is therefore the most promising target and, if a successful sighting of this moon is made, the next most likely candidate is probably Europa, a very interesting moon with a possible subterranean ocean. It is both fainter and closer in than Ganymede, however, so finding it will be a more difficult feat. Good luck!

Telescopically, the satellites are very conspicuous, and tracing their dance around their giant primary is fascinating. Sometimes a moon will pass in front of Jupiter in transit. Not only is the moon itself silhouetted against the cloudy surface of the planet, but its black shadow on the clouds below can be surprisingly conspicuous. Indeed, the shadow is easier to see than the moon itself, and the present writer recalls seeing one such event in a 2.5-in. (6-cm) refracting telescope. On other occasions, a moon will pass behind Jupiter, and for a while remain hidden (or "occulted," to use the technical term) by the giant world. The moons can even pass in front of each other, as well as being eclipsed by the shadow of the planet. These latter events are quite fascinating as the moon just fades out while clear of the planet's limb. A good pair of binoculars can easily catch such an event; now you see it ... now you don't!

As for the moons themselves, although bright in even a small telescope or moderate pair of binoculars, they are usually described at being star-like in these instruments. Nevertheless, there have been reports that, under ideal conditions, telescopes as small as 6 in. (15 cm) have been able to see them as tiny discs, and it has even been claimed that a 4-in. (10-cm) telescope revealed not just their discs, but also markings (!) on at least one occasion. Claims of this sort are probably best treated with a degree of caution, as marginal "details" near the limit of observation can easily be subjective. Even under very good conditions, observers using telescopes of 12 in. (30 cm) in diameter find it difficult to detect markings on the brighter satellites of Jupiter.

Some Interesting Pre-Discovery Observations

The actual "discovery" of an astronomical object does not necessarily mark its first observation. From the former item, we see that Galileo was probably not the first person to see the bright moons of Jupiter. The great importance of his discovery was not so much in its being the first sighting of the moons (which it seems not to have been), but in Galileo's realization of their significance and the integration of this into the body of human knowledge. This is the mark of true scientific discovery, and this is not taken away from the discoverer if it is found that he/she was not the first to spy the object in question.

Indeed, it is not in the least surprising that relatively bright objects such as the planet Uranus did not wait for their official discovery before being noted by casual observers. This planet, it would seem, was actually seen on at least 23 occasions between 1690 and the date of its official discovery by W. Herschel in 1781. Even John Flamsteed, England's first Astronomer Royal, noted a "star" on the night of December 3, 1714, which turned out to have been none other than Uranus. It would not be surprising if other early observations of Uranus (and also of the bright asteroid Vesta that, on rare occasions, can indeed outshine it) lay gathering dust in some musty basement.

Neptune, though fainter than either Uranus or Vesta, is another object unlikely to have awaited its official recognition. As a matter of fact, we know of one pre-discovery observation of this planet, by none other than Galileo himself!

Thanks to research by Charles Kowal and Stillman Drake, this distant planet can be identified as one of the "stars" drawn by Galileo in his representation of the field of Jupiter on January 28, 1613. We can only speculate what would have happened if this great astronomer had noticed that one of the supposedly fixed stars in his telescope field was moving. Would he have recognized it as a planet? In that case, Neptune would have been announced before Uranus and the history of our knowledge of the outer Solar System taken an entirely different path.

Sunspots constitute another type of phenomenon for which early pre-discovery observations exist. They were noted in early times by Chinese astronomers-astrologers, and, according to Xu Zhen-Tao, records even exist in the *Book of Changes*, compiled some time before 800 BC!

Sunspots large enough to be visible with the naked eye are, in fact, quite commonplace, but the great brilliance of the Sun coupled with the danger to eyesight that any unprotected solar observation involves, makes their sighting a far from easy task. Advanced though the ancient Chinese were, they did not have solar filters! So how did they make their observations without going blind in the process? (One may indeed wonder how many *did* ruin their eyesight before the dangers of solar observing became fully apparent.) Did they discover the Sun's blemishes while observing its image reflected in pools of water? Were the first observations made when the Sun was dimmed by low altitude and a heavy pall of dust or cloud? We can only guess.

Daylight Observing

There is more to be seen in the daytime sky besides the Sun and Moon. Believe it or not, two planets and at least two stars have also been spotted by naked-eye observers, and quite a few other stars are within daytime telescopic range.

Of the planets, Venus is the most prominent daylight object. The brightest permanent resident of the sky other than the Sun and the Moon, this queen of planets is observable most days with unaided eyes. The biggest difficulty is locating the tiny pinpoint of light in such an expansive blue desert, but once it is found, it is remarkably easy to see. The best chance is when the crescent Moon is also visible close to Venus. The former acts as a pointer to the planet and also sets one's eyes at the correct focus for distant objects. Another way of finding it in daylight is to start watching before sunrise, when the planet is in the morning sky and keeping it in view as the Sun comes over the horizon.

Nevertheless, because it appears so small in a very large and mostly empty sky, not many people see it and, if it happens to be located by accident, the person seeing it may be very incredulous that the object of his attention really is Venus.

Back in 1970 or thereabouts, a major UFO scare was triggered in a regional city in New South Wales when, day after day, a bright silvery object appeared in the daylight sky. After a few days, crowds gathered at the prime viewing site to watch the "phenomenon," and for a time, speculation ran riot. Somebody would shout, "There it is" and (with gasps of wonder) dozens (probably hundreds after a while) of people turned eyes, cameras, and binoculars skyward in anticipation. The "UFO" was even televised on the national news broadcasts!

This lasted until somebody plotted the object's position with enough precision to allow the government astronomer, Dr. Harley Wood, to positively identify it as Venus, after which the whole episode quietly faded from view like a setting planet.

Long before this event – as far back as February 1, 1798, in fact – a crowd gathered outside Luxembourg Palace to applaud Napoleon on his return from the victorious Italian campaign, allegedly became distracted by the appearance of the planet in the noonday sky. One wonders what Napoleon thought of that!

The writer has seen Venus on numerous occasions with the naked eye in full daylight. This author can even recall – quite a number of years ago now – watching an occultation of Venus by the Moon through an office window in the middle of the afternoon.

Yet, I have never been able to spy Jupiter in broad daylight without a pair of binoculars. With binoculars it is easy, once the exact location is known, and I was impressed as to what a great "flying saucer" the planet made in a pair of 20×65 s. It really *did* look like a tiny silvery disk!

Others have had better luck finding the king of planets in the daytime without optical aid. Writing in the December 1976 issue of *Sky & Telescope*, astronomer and author Fred Schaaf reported three such sightings by himself and two other people earlier that same year.

The first was by Steve Albers on July 21, 1976. While camping under the clear skies of Zion National Park in Utah, Albers noted that Jupiter was quite close to the Moon in the pre-dawn sky and figured that if it was possible to see the planet in daylight, this would be a golden opportunity. Using the Moon as a reference object, he was able to keep Jupiter in naked-eye view until 10:21 a.m. local time, by which time the Sun was high up.

Noting that a similar opportunity was predicted for August 18, Schaaf drew attention to the possibility of a daylight sighting through his astronomical column in the Atlantic City newspaper *The Press*. The announcement proved fruitful with two naked-eye sightings reported, one by Schaaf himself. Schaaf's own sighting was rather brief, as he was only able to hold the planet in view for a little while after sunrise, but he later received a letter from a reader of his column (Chuck Fuller) who managed to follow it until 11:15 that same morning.

Scaaf's observation is also remarkable for a different reason. As well as watching Jupiter with his unaided eyes that morning, he also followed it with a 4.3-in. (11-cm) reflecting telescope for some 20 min into daylight. That in itself may not seem especially remarkable, but in addition to Jupiter, he managed to hold the brightest moon Ganymede in view for the same length of time!

A similar daylight observation of Jupiter and its largest moon was reported some 3 years earlier by E. G. Moore of London, England. On June 22, 1973, Moore used an 8.6-in. (22-cm) reflector to follow Jupiter's moon for 30 min after the Sun had risen. He kept the planet itself in view for a further 23 min.

The appearance of both Jupiter and Venus together in the morning sky was responsible for a UFO scare involving two police officers on early morning patrol. Probably tired and fatigued, the officers became convinced that the two "lights" were chasing them, and the whole incident got out of hand, as such panics are apt to do. The rather embarrassing story need not be repeated here, but the interesting feature, for our purpose, is the officers' insistence that *both* the "lights" remained visible after the Sun was up. Perhaps too much should not be made of this, given the circumstances of the sighting, but it implies that Jupiter must have been relatively conspicuous for at least a brief while after sunrise.

It is most likely that the other planets have not been observed with the naked eye during daylight hours, although Mercury, Mars, and Saturn are well within daytime telescopic range. In fact, some observers have found daytime observations of Mercury more reliable than evening or morning ones, as the planet rides high during daylight hours, relatively free from the distorting effects of our atmosphere. Mars, it should be noted, occasionally outshines Jupiter and so should theoretically be within daytime naked-eye range. The catch is, this only happens near opposition (only near very favorable oppositions, in fact) when the planet rises around sunset and only reaches observable altitude in a dark sky.

Turning to the stars, the one most commonly observed in daylight without optical aid (other than the Sun, of course!) is Sirius, but the bright southern luminary Canopus has also been located while the Sun was above the horizon. One morning in 1985, Rob McNaught of Siding Spring Observatory in Australia followed it with his unassisted eyes for a short while after sunrise. Apparently the ancient Chinese managed Arcturus during daylight hours, although there are no reports of anyone recently achieving this feat (although there have been telescopic sightings). If naked-eye sightings of this star are possible, however, Alpha Centauri should also be within range. Southern hemisphere readers may like to try for this star, and readers from either hemisphere might like to try for the slightly fainter Arcturus.

When the eye is supplemented with a telescope, it is difficult to know where the limit of daylight visibility can be drawn. For example, the same E. G. Moore who saw Jupiter and Ganymede by day also located Polaris (at roughly second magnitude) 1 h after sunrise with a 9-in. (23-cm) short-focus refractor at 50 magnification. And that, despite a dewed lens!

On July 7, 1959, a very interesting and rare event occurred. The planet Venus eclipsed (or, more correctly speaking, occulted) the 1.4 magnitude star Regulus. From England, this event occurred in the early afternoon, yet was well observed by astronomers Henry Brinton and Patrick Moore with the aid of Brinton's 12.5-in. (32-cm) reflecting telescope. These observers noted that the star perceptibly dimmed for about one second as its light passed through increasing depths of the Venusian atmosphere, before blinking out completely.

Actually, it is not uncommon to find, in the notes of some nineteenth century astronomers such as Dawes, references to stars as faint as third or fourth magnitude being found during sunlit hours with relatively small instruments. It seems that stars may be easier to see in daylight than is usually believed. The biggest problem may be finding their location and making sure that the telescope is in exact focus.

While on the subject of daylight star sightings, there is a curious story that many readers will probably have heard, that tells of stars having been seen in daylight from the bottom of deep wells and the like. Not far from the writer's home are some large electricity generating plants with huge smokestacks, and I recall that, during the construction of the first of these, a tale circulated that if the workers looked up through one of the stacks, stars could be seen in full daylight. This, I stress, was a story in circulation. I did not hear it directly from anyone who claimed to have had this experience.

What are we to make of this?

For a start, we admit that if Venus, Jupiter, or (perhaps) Sirius passed directly over the smokestack, somebody inside the stack and looking skyward might very well see it. The drawback to this, however, is the sheer improbability that one of these bright objects would just happen to be present in the tiny circle of sky visible through the opening at the top of the stack. This would surely be equivalent to firing a random shot into the sky and bringing down a duck!

So does observing through a smokestack or from down a well actually darken the sky to the extent that "average" stars can be seen during the daytime?

It would seem not. Experiments show that the sky seen through the small opening remains bright. Excepting the very unlikely fluke just mentioned, stars should not be seen through a smokestack or from down a well!

But if that is true, what are we to make of the story? Is it just another urban legend or a led-pull that has managed to become part of folk belief?

Maybe not. It is possible that bright points of light have been seen under these circumstances. The problem is they are not stars! Smokestacks create updrafts, and insects may fly around the mouths of wells. Could it be that the triggers for these reports have been insects or tiny pieces of debris lit by the Sun and (at least in some instances) held aloft by updrafts? It would be interesting to question anyone who actually claims to have seen these "stars" with his or her own eyes (as against someone who claims that a friend told him about a friend of a friend ...!) as to whether and in what direction they appeared to move.

Project 21
Daylight Observing

What is the brightness of the faintest star that can be seen in your telescope when the Sun is above the horizon?

Finding a star in daylight is no easy task. The telescope must be positioned exactly, but the focus must also be exactly right, or the point-like star becomes a fuzzy disk that easily melts away into the bright background. If the Moon or Venus is visible, getting a sharp image of these should fix the focus problem, but precisely locating a specific star still takes a lot of exact work with setting circles; digital or otherwise.

The best way of locating stars in the daytime is to focus on them while the sky is still relatively dark, prior to sunrise, and hold them in view until the Sun appears over the horizon. A steady telescope is essential for this, and a drive that keeps the star in the center of the field of vision is highly desirable.

First of all, try some easy objects such as Venus and Jupiter, then move on to bright stars such as Sirius before trying for the more challenging objects. For northern observers, a good object in the fainter category is Polaris as, although far from brilliant, being so close to the celestial pole it is at least slow moving!

Fainter objects in close proximity to bright ones are also good for a challenge. How many of the Galilean moons of Jupiter can you hold in vision after sunrise – and for how long? By picking up Jupiter later in the day, is it possible to find any of the moons when the Sun is high?

Now that is a *real* challenge!!

Visibility of Faint Stars

Most basic books on astronomy give six as the faintest magnitude discernible with the naked eye on a clear dark night. This is a fairly good average for good eyes and good skies, but the practical limit will, of course, vary significantly according to an observer's location. The artificial lighting of a great city makes a huge difference to what one can see with the unaided eye. Stars that are easily visible without optical aid from the countryside are reduced to binocular visibility close to a large urban center. Nebulous objects such as the Andromeda Galaxy and the brightest globular clusters – not to mention the Milky Way band itself – simply blend to invisibility into the light background sky. Amateur astronomer Joseph Rao recalled that he had not seen the Milky Way from his home in the Bronx for 16 years, until the night of July 13, 1977, when a power failure plunged the city into darkness!

At the other end of the scale, under skies that are clear, dark, and pristine, naked eyes frequently reach deeper than magnitude six. The present writer has seen down to magnitude seven or slightly fainter with the unaided eye (well, unaided except for eyeglasses!) from rural skies in central New South Wales. On an unusually dark night at Matecumbe Key, Florida in 1973, Martin Hale (down from Canisteo in New York) managed to espy stars as faint as magnitude 7.5 as well as counting ten naked-eye stars in the Pleiades – the so-called "Seven" Sisters cluster. Hale's feat is not the record for this star group, however. Amateur astronomer Ronald Pegram of Greensboro picked up 15 Pleiades with eyes alone on one exceptionally cold and clear night. At that time, he was presumably seeing to even fainter magnitudes than Hale's 7.5.

For seeing *really* faint stars, nothing beats going to the top of a very high mountain (unless one ventures into space, of course!) such as Mauna Kea in Hawaii. In January 1985, keen-sighted astronomer Steve O'Meara glimpsed stars slightly fainter than eighth magnitude without optical aid and, peering through a 24-in. (61-cm) telescope at the Mauna Kea Observatory, reported finding stars of between magnitudes 19 and 20. On that night, he even managed to sight the approaching Halley's Comet, just over 1 year before its 1986 rendezvous with the Sun. At that time, the comet was just 19th magnitude, and the initial report understandably met with some degree of skepticism from a number of astronomers, although this attitude quickly faded once a more detailed account of O'Meara's observation became known.

...And a "Bright" Quasar!

Quasars, as every astronomer knows, are exceedingly remote objects. Although intrinsically of truly awesome brilliance, they are so far away that most require large telescopes and/or CCD technology to find them.

There is, however, one exception. The very first quasar discovered, 3C 273 in the constellation of Virgo, can be seen with backyard telescopes. The writer once hunted it down just for the sheer novelty of seeing something 2 billion light years away. It is both humbling and strangely weird to think that the photons entering your eye left their source at a time when nothing more complex than blue-green algae lived on Earth and free oxygen was only starting to become dominant in our planet's atmosphere.

Nevertheless, the quasar was, if not "bright" in the usual understanding of the word, at least obvious in the 10-in. (25-cm) reflecting telescope being used at this time. It was obvious that smaller instruments could still have reached it. The question is, "How much smaller?"

Many astronomy texts inform us that the quasar is visible in 8-in. (20-cm) telescopes. The object is actually slightly variable, with magnitudes ranging unpredictably between magnitudes 12 and 13. The writer has detected stars around these magnitudes

Quasar 3C 273, the brightest in the heavens. Courtesy J. Bachall (IAS). NASA image.

(especially toward the brighter end of the range) with a 6-in. (15-cm) telescope, so the quasar should be accessible to instruments of that size on some occasions at least.

Apparently, though, a 6-in. 'scope does not even come close to the smallest instrument capable of reaching this quasar. Under clear Arizona skies, astronomer Brian Skiff managed to find it using only a 2.8-in. (7.1-cm) refracting telescope and, upon reading of this feat, husband and wife observers Alan and Sue French tracked it down under less favorable skies with the aid of a 3.6-in. (9.1-cm) telescope at 114 magnification (Sue) and a 4.1-in. (10.4-cm) at a magnification of 87 times (Alan). Probably the most remarkable sighting of 3C 273 thus far, however, was accomplished by Finnish amateur astronomer Jaakko Saloranta, who spied the quasar through a refracting telescope with an aperture stopped down to just 1.6 in. or 4.0 cm! The magnification being used at the time is not known to the writer, but it had to be about as high as this very small aperture could take.

Project 22
Can 3C 273 Be Found Using
Large Binoculars?

Most would answer "No" to this question straightaway, probably followed by gales of laughter that anyone would ask it in the first place!

Nevertheless, it is worth noting that under clear and dark skies, large high-magnification binoculars such as 25 × 100s can detect stars of magnitude 12 and fainter, and there seems no reason to think that 3C 273 could not be glimpsed during its brighter phases using such an instrument. Perhaps the reader might like to try for it. Although a scientifically useless pursuit, finding the brightest quasar with a pair of binoculars would nevertheless be very satisfying. Just one tip though. Only admit that you tried if you are successful. That way you will become known for your observational skills and not for being odd!

Very Thin Crescent Moons

How close to new Moon can the thin crescent of the Moon be seen?

Certainly within a single day, if circumstances are just right. Several sightings around 20–21 h from new are on record, with a few not much greater than 19 h from the moment of "no Moon."

Sightings even closer to new have been reported on rare occasions, although these are extremely difficult. According to A. D. Thackeray, the sighting of an ultra-thin crescent by Dr. Harold Knox-Shaw, a mere 17 h and 30 min from the moment of new Moon set the record for his day. This "record" was, however, broken in 1910 when a 16-h old crescent was sighted by D. W. Horner of Kent on February 10. The latter was almost equaled on August 13, 1931, when famous lunar expert Andre Danjon spotted an extremely thin waning crescent just 16 h and 12 min before new, through a 3-in. (7.5-cm) refracting telescope.

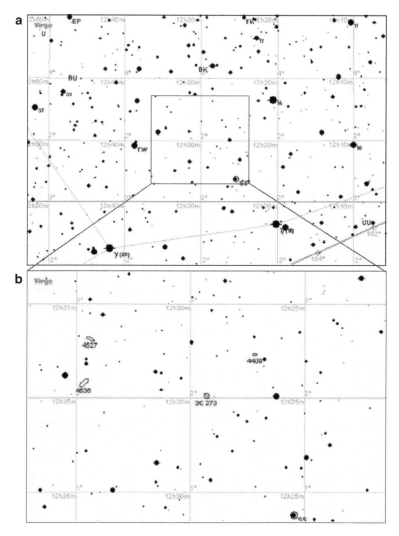

Finder chart for the sky's brightest quasar. Top chart shows stars to magnitude 8. Lower chart shows stars to magnitude 13, approximately that if the quasar itself. © Emil Neata, used with permission.

More recently, in May 1990, that master of difficult observations, Steve O'Meara, spotted a very thin crescent with his unaided eye, just 15 h and 32 min after new. Even that was not a record, apparently, as a little more digging will reveal that a crescent was seen just 14 h and 45 min from new by Mr. Hoare of Feversham (also in Kent) on July 22, 1895 (probably with optical aid). This feat came close to being matched on March 15, 1972, when Robert

Moran located a very thin crescent by using 10×50 binoculars just 14 h and 53 min from the exact time of new Moon.

Amazingly, still thinner crescents have been spied from time to time. On May 2, 1916, Lizzie King and Nellie Collinson managed to find the Moon just 14 h and 30 min from new. Even this pales in comparison with an observation of what is believed to be the youngest crescent Moon ever seen with optical aid. This took place on September 7, 2002, when Mohsen G. Mirsaeed of Teheran saw a crescent just 11 h and 40 min after new! This record is unlikely to be broken, unless observers start counting seconds and fractions of seconds on a stopwatch. Maybe not even then!

During the 1930s, A. Danjon calculated that when the Moon comes within about 7.5° of the Sun, no illuminated crescent is possible. While within that region of sky, the Moon is invisible from Earth, unless it actually passes in front of the Sun. But that is a different story!

This minimum elongation is known as the Danjon limit, and, although widely quoted, it is not immediately clear whether it measures an intrinsic property of the crescent Moon or simply the human eye's ability to see very thin threads of light. Recent findings apparently confirm the latter as we shall see in a moment. But as we are currently concerned only with the thinnest crescents that we can *see* rather than with those that might be detectable by some means other than normal vision, we let the Danjon limit hold.

As the mean hourly motion of the Moon against the background sky is around 0.55°, the distance equal to the Danjon limit is covered in 13.6 h. When the Moon passes directly in front of the Sun in a solar eclipse, it remains invisible for at least that amount of time afterwards. So the Mirsaeed record won't be challenged during eclipse new Moons!

Returning to a point made above, we again raise the question as to whether the Danjon limit necessarily holds for all types of observation, including those made photographically or in wavelengths other than the visual, or whether it is relevant only to observations (whether naked-eye or telescopic) made visually.

The question really hinges on the true nature of the Danjon limit, viz., whether it is truly a limit as to how thin a crescent can become or simply a limit as to how thin our eyes can perceive one.

Early observations from rockets shot above Earth's atmosphere appear to have placed the limit squarely in the field of human perception. Extremely thin crescents were photographed from these when the Moon was just 2° from the Sun!

More recently, Martin Elsasser and colleagues conducted a series of experiments in imaging extremely thin lunar crescents during broad daylight from very favorable sites in the Austrian mountains. Using a baffle to block direct sunlight from the telescope, Elsasser et. al. imaged the Moon at both visual and infrared wavelengths to observe crescents far closer to the Sun than any previous Earth-bound observers had managed to record. Using this method, they managed to image (but not to "see") the Moon just 4.58° from the Sun, only minutes from conjunction on May 5, 2008. Remarkably, the Moon can be followed right through the New phase, at least during certain months when minimum elongation does not reach the smallest possible angles.

Elsasser writes that, "Due to the numerous technical means required to capture this faint lunar crescent, these images probably do *not* show what could be seen through a telescope. It is quite

Very slender crescent Moon, just 3.5 h from new! Courtesy, Martin Elsasser.

Not easy to see, this photograph includes an image of the youngest lunar crescent ever observed, literally minutes away from new. The crescent may be traced as a very thin thread in the lower left of the image. The brightest section of the crescent is directly below center, almost at the bottom of the frame. Courtesy, Martin Elsasser.

possible that the lunar crescent will never be visible to human eyes from earth at these elongations, due to low contrast. Of course, the lunar crescent is still there, even if we cannot see it with the naked eye. The main reason for using cameras is to capture things the eye could not see and to document these."

In other words, although the Danjon limit appears to stand for visual observation, Elsasser's work clearly demonstrates that it breaks down when the human eye is replaced by a camera!

Project 23
Seeking the Thinnest Crescents

To have any chance of seeing the very thinnest of crescent Moons, everything must work together to ensure that the Moon's angular separation from the Sun is as great as possible. Look for them when

(a) The Moon is at or near perigee (i.e., its closest point to Earth), as at these times it is moving faster than average against the background sky

(b) It is near greatest ecliptic latitude (=5.5 or −5.5), adding to its angular distance from the Sun

(c) It is close to the Sun's azimuth, ensuring that it is almost directly above the setting Sun (or rising Sun for "old Moon" crescents). This gives maximum height above the horizon in as dark a sky as these small angular distances allow

In addition to these factors, anyone hunting very early – or very late – crescent Moons will also find it of benefit to gain as much altitude as possible in the mountains, where skies are clearer and darker, and where much of the haze and humidity of lower altitudes is left behind.

Finding a crescent Moon within 15–16 h of new is truly a remarkable feat and must surely rank high among the *really* difficult observations. It may not count as a scientific breakthrough, but it certainly brings personal satisfaction!

Rogue Tales and Observations

The stories in previous chapter are interesting, not so much for the objects observed as for the nature of the observations themselves and the conditions under which they were made.

Not so with the tales we are now about to recount. If any of these purported discoveries could be confirmed, our knowledge of the universe would be changed, at least a little, and we would need to review our attitude toward certain matters about which we are now skeptical. It should be remarked right here that, although some of the alleged observations about which we will be speaking have been proven incorrect (and others *almost* proven incorrect), the door has not been shut on all of them. Some could yet become the orthodoxy of tomorrow!

The Strange Stories of Gelatinous Meteorites

Imagine that you are out of doors at night in the country. It is a beautifully clear and dark evening with stars blazing like diamonds against the velvet black of the sky. Suddenly, your attention is arrested by a beautiful meteor blazing through the heavens. Brighter than a normal shooting star, the meteor dives downward toward the ground, appearing so close that you could imagine it falling in the open field just a few minutes walk away. Then, as suddenly as it appeared, it is gone. All that remains is a quickly fading phosphorescent streak in the heavens marking its flight.

You stand awestruck as this ghost of the fireball fades away when, unexpectedly, the silence of night is interrupted by a dull thud, seeming to originate in the direction of the meteor fall. The night being so dark, you do not investigate immediately, but carefully mark the direction and, come daylight, walk over to have a closer look. To your surprise, you find, lying in the open field, a fetid-smelling jelly-like mass!

This sounds like something from *The X-Files*, but believe it or not, it is based on actual accounts of a phenomenon that (if it really exists) is very weird indeed.

As long ago as 1638, we find an account of a church at Dartmoor in England being struck by a bolt of lightning (?) that was reported to leave behind it a foul-smelling gelatinous mass. Sections of the account are worth quoting, both for the details of the event itself and to savor the language style of the period.

Following the event, according to this record, the church was filled with a "loathsome odor." The account continues: "It seems that some person who ventured to go up into the wrecked tower, there discovered 'a round patch as broad as a bushel, which looked thick, slimy, and black, to which he put his hand, and found it soft, and bringing some from the wall, came down and showed that strange compound. It was like a slimy powder, tempered with water; he smelling thereto, it was odious beyond expression, and in a far higher degree of loathsomeness than the scent which was in the church when they first smelt it, being of the same kind, and the discoverer was shortly after attacked with severe colic." Despite, or maybe because of, the sometimes quaint language, the point is most aptly made!

Again, on March 24, 1718, gelatinous matter accompanied by a ball of fire was seen to fall on the Isle of Lethy in India, and toward the end of the same century (on March 8, 1796) "viscous matter" fell with a meteor at Lusatia, Europe. The "viscous matter" referred to was said to have possessed the color and odor of varnish.

A report from Silesia in Germany recounts that on January 21, 1803, a meteor fell with a "whizzing sound" and was said to have remained "burning on the ground" (this is very curious, as the fireballs accompanying meteorites typically burn out several miles above the ground, as we have already remarked). In any case, according to this report, a "jelly-like mass" was found on the snow the following day, apparently at the exact position of the "burning meteor."

Just briefly glancing at a cross section of other reports, we note a meteor falling "twenty yards off" from an observer on October 8, 1844, and being heard to strike the ground with a noise. Next morning, a grayish gelatinous mass so viscid as to "tremble all over" when poked with a stick was found at the spot of the alleged fall.

A little over 2 years later, a meteor as bright as the Sun seen from Loweville, New York, was said to have fallen in a field. Upon seeing this, "a large company of the citizens immediately repaired to the spot and found a body of fetid jelly, four feet in diameter."

There are many similar accounts, but these are sufficient to give us an idea of the phenomenon.

A constant feature of these stories, it will be noted, is the assumption that the meteorites fall only a short distance away. This is dubious; *very* dubious! If a meteor appears to burn all the way to the ground, you can be sure that it has actually sunk beneath the horizon and is really a long way off. (This does not apply to a giant crater-forming meteorite, but if you happened to be near the fall of one of these, misjudging distance would be the least of your concerns!). It is well known that meteorites always appear much closer than they are. Meteor experts say that quite frequently a witness will be convinced that a meteorite fell "in the next field" or the like, when it may actually have been tens of miles distant. On these grounds alone, a real association between a meteor and a gelatinous mass on the ground is doubtful. If the meteorite fell many miles away, it is hardly blameworthy for a heap of foul-smelling gunk in the next-door field!

In fact, many of the meteors mentioned in these accounts simply did not seem the kind capable of dropping meteorites. A meteorite fall is accompanied by some pretty spectacular effects – not only a brilliant fireball but also sonic booms that can be quite terrifying if the fall is nearby. These grand displays are missing from most of the "gelatinous meteorite" accounts.

Most probably, the gelatinous masses were merely rotting organic matter with a purely terrestrial origin. The meteor was simply a coincidence. Although some of the reports – such as the very early account from Dartmoor – do not fit easily into this scenario, it at least has the advantage of being the simplest explanation unless some indisputable contrary facts come to light.

Red Is the Rain that Falls on Kerala

Gelatinous material is not the only organic stuff suspected by some of falling from space. Between July 25 and September 23, 2001, the southern Indian state of Kerala witnessed a most unusual phenomenon – red rain! The rain came in a series of heavy downpours, and anyone caught in them ended up wearing clothes stained the color of blood. Other colors of rain (yellow, green, and black) were also reported, but the red coloration seems to have been the most widespread.

Actually, this was not the first time that red rain had been reported in Kerala. There was a fall as far back as 1896, and on several other occasions during the intervening years, but none engendered such widespread attention as the falls of 2001.

There may be a certain irony about a state with an elected communist government experiencing red rain, but whatever the explanation is, it is certainly not political!

One would hardly assume it to be astronomical either, which should place it outside the scope of this chapter. It is included, however, on the grounds of two astronomical associations that were made by early investigators of the phenomenon.

Initially, a story circulated that an exploding meteorite preceeded the first fall of the recent series. According to locals in the areas initially experiencing the rain, that fall occurred following a "loud thunderclap and flash of light," after which groves of trees began to shed "burnt" leaves. Reports also recounted the sudden formation of certain "wells" and the disappearance of others in the region of the event.

The second astronomical association grew out of the first, but more about this in a moment. First, though, a few words must be said as to how the rain acquired its red coloration

The rain did not contain *dissolved* red material. Its color was due entirely to *suspended* red particles. Indeed, if a glass of the red rainwater was allowed to stand for a time, the red particles settled out at the bottom, leaving clear water in the rest of the vessel.

It was determined that each milliliter of rainwater contained some nine million red particles in suspension and that each liter of the water contained around 100 mg of solids. Clearly, whatever the particles were, they existed in enormous numbers.

Worldwide interest in, and controversy about, the phenomenon broke in 2006 with the announcement by Godfry Louis and Santhosh Kumar of the Mahatma Gandhi University in Kottayam that the red particles were in fact tiny organisms of a completely unknown type. In their on-line report, these scientists made a claim to have isolated one of the microorganisms and found it to show "very extraordinary characteristics like [the] ability to grow optimally at [572°F (300°C)] and the capacity to metabolize a wide range of organic and inorganic materials." Curiously, after analyzing the organisms using ethidium bromide, these researchers failed to detect the presence of either DNA or RNA, which they interpreted as evidence that these allegedly living particles represented a form of life unlike anything else on planet Earth – life that was not based on nucleic acids. Not, at least, on the ones we know.

> It's life Jim, but not as we know it!!!

Harking back to the possible exploding meteorite, these researchers concluded that the microorganisms must have come from space!

According to their radical and controversial suggestion, a meteorite laden with alien microorganisms exploded over Kerala state, spreading its biological load far and wide through the atmosphere. The alien life forms were eventually washed down in rain, giving it the strange red coloration.

The thought of aliens (even if only microscopic ones) falling to Earth made good news copy, of course. Most researchers, however, remained unconvinced.

By contrast with the Louis/Kumar results, analysis by Chandra Wickramasinghe at Cardiff University showed positive for DNA. If confirmed, this probably weakens the extraterrestrial claim. Moreover, Milton Wainwright of Sheffield University found similarities between the red particles and algal spore, also implying a terrestrial origin. Similar conclusions were drawn from an independent study commissioned by the Indian government.

On another level, the earlier reports of red rain in the same general region must also arouse suspicion as to the validity of the alien organism suggestion. Although we cannot be certain that all of these events were caused by the same thing, that is at least a good assumption in the absence of evidence to the contrary. Surely, it is too much to believe that rare alien-bearing meteorites have been sporadically exploding over Kerala ever since 1896! Some local source for the particles is certainly more feasible.

Moreover, the alleged meteorite supposed to have preceded the 2001 rains rests on very scant evidence. Reports of flashes and bangs do not necessarily add up to a meteorite fall. Maybe there were some unusually severe lightning bolts before the rain. When all is said and done, no indisputable sightings of a meteor or meteorite exist, which is hard to explain if such a spectacular event really did occur.

With these final thoughts, any purported astronomical association with the Kerala red rain unravels. As such, the story is no longer relevant to us and we are probably justified in leaving it to the biologists and meteorologists to work out in finer detail!

Strange Structures Found in Meteorites

Ever since meteorites were first recognized as having fallen from space, they have been imbued with both the mystique and value due to them as pieces of extraterrestrial material. Indeed, until the first Moon rocks were brought back on board *Apollo 11* and its successors, they represented the only sizable chunks of material from beyond Earth that human beings could handle and analyze. Many clues about the formation of our Solar System and, by extension, our home planet have been gleaned from these extraterrestrial arrivals.

Lessons on how the Solar System formed were highly prized, but there was also another hope lurking in the background. Could meteorites yield up some clue about the possibility of life beyond our planet?

Something of the kind was in the mind of the unknown wag who deliberately planted biological material in a fragment of the famous meteorite that fell at Orgueil in France on May 14, 1864.

This body was a member of a type known as CI meteorites or Type 1 carbonaceous chondrites. As their name implies, they contain a significant amount of carbon, much of it in the form of relatively complex organic materials, but most of their substance is a clay-like material that is rapidly eroded and even becomes soft and malleable when wetted. A piece of meteorite of this class will actually dissolve (or at least turn into a layer of sediment) if dropped into a container of water.

Well, the unknown joker wet a piece of the Orgueil meteorite to make it soft, and then impregnated it with seeds and other pieces of obviously biological material. He did not get the last laugh, however, as the fraud was not found out until about a century later. Nevertheless, the very fact that a meteorite fragment could be deliberately contaminated in this way stands as a demonstration as to how one may inadvertently become contaminated and how we must be very careful before assuming that everything found encased within a meteorite actually arrived with it from the depths of space! We will return to this in a little while.

It is also true that some features in meteorites, or any rock for that matter, can imitate fossil remnants of living organisms and yet be totally inorganic in their true nature. During the second half of the nineteenth century, German scientists Drs. Hahn and Weinland examined thin slices of stony meteorites and found tiny structures that they identified as fossilized remains of sponges, corals, and crinoids. Weinland even went so far as to classify a number of the "corals" as belonging to the extinct class of favositines. One alleged coral was named by Weinland *Hahnia meteoriticia* in honor of Dr. Hahn, whose analysis first uncovered the structures.

Although the apparent fossils appeared to be morphologically similar to their terrestrial counterparts, their comparative sizes differed greatly. All of the alleged fossils measured less than one millimeter in diameter. Most were considerably smaller than this and could only be seen to advantage with the aid of a powerful microscope.

Based on the results of these scientists, Francis Birgham wrote in an article in *Popular Science* in 1881 that the existence of living organisms of these varieties implied that the parent body from which the meteorites originated was of planetary dimensions and possessed liquid water. The thesis for which Birgham opted was that of a planet that was smashed to pieces some time in the remote past, though not before an evolutionary process parallel to that of Earth had already begun.

Alas, the results of Hahn and Weinland did not withstand the tests of time. Rather than fossils of Lilliputian corals, sponges, and the like, these meteorite structures were simply rock crystals and similar mineral structures viewed with an enthusiastic eye.

The claims of Hahn and Weinland nevertheless paled in comparison with those made in the 1930s by Professor Charles B. Lipman. As part of his study of bacteria in rocks, Lipman claimed to have cultured living organisms not only from ancient terrestrial coal deposits but from meteorite specimens as well. Arguing that the presence of these organisms was not due to contamination, Lipman asserted he had grown cultures of both ancient microorganisms and of extraterrestrial life.

Needless to say, these claims did not remain unchallenged for long.

Thus, Sharat K. Roy repeated Lipman's meteorite experiments with negative results. Admittedly, cultures did grow, but they consisted only of the sorts of micro-organisms expected to be found lurking in a laboratory environment. Moreover, the control cultures seemed to grow as well as those from meteorite fragments!

Lipman vigorously criticized Roy's work, but once again the more sensationalist claims failed to be upheld by future research, and it is virtually certain that nothing more exotic than contamination was responsible for Lipman's results.

The trouble is, Earth is so filled with life and the products of life that contamination is very, very, difficult to avoid. This is especially true of very porous meteorites such as Type 1 carbonaceous chondrites. Recall the ease with which some joker contaminated a piece of the Orgueil meteorite. Nature can perform the same hoax just as readily. Pores in meteorites share the vacuum of space and fill rapidly as the body enters Earth's atmosphere. Microorganisms and spore have been found at very high altitudes and are easily taken into the meteorite as the vacuous pores fill with air.

This issue of contamination raised its head again in the third quarter of the previous century with the finding of "organized elements" by B. Nagy and colleagues in carbonaceous chondrites – once again, most notably the Orgueil. These "elements" appeared as very regular inclusions and often looked remarkably like fossils of simple forms of life. Their discovery caused quite a stir at the time, but since the identification of some of the inclusions with terrestrial pollen, the impression has circulated that they have all been dismissed as contamination. That is not really true, as several different types of organized elements were described by Nagy, not all of which can be identified with contaminants and not all of which have been satisfactorily explained.

Even harder to dismiss are the ultra-fine structures found on Professor H-D Pflug's transmission electron micrographs of thin sections of the Orgueil, Murchison, and Allende meteorites. These meteorites represent the three most populous categories of carbonaceous chondrite, Types 1, 2, and 3 (or, to be more precise, types CI, CM, and CV). Both the Murchison and Allende meteorites fell in 1969, in Victoria (Australia) and Mexico, respectively.

Most of Pflug's samples were of the Murchison and revealed micron and sub-micron structures not unlike the organized elements of Nagy, only far smaller and constituting a significant portion of the organic content of the meteorite. Many revealed a clear bilaminar membrane structure similar to elementary cells, and there was even evidence of these microvesicles having multiplied through budding. Some appear to have grown into relatively large colonies through this process.

Interestingly, the membranes of the microvesicles were surrounded by a sort of covering that Pflug suggested may have arisen as the excretory product of a simple metabolism within the vesicle itself. He did not use the word "life" in connection with this process, and unless the definition of that term is extended somewhat, to refer to it as such would probably be an exaggeration. If his interpretation of a simple metabolism is correct, the best that it could be termed is probably "proto-life."

Pflug suggested that chemicals were absorbed by the microvesicles through the bilaminar membrane and concentrated inside the vesicle, where comparatively complex reactions take place and more complex organic molecules are synthesized. In effect, the vesicle "fed" on simple chemicals through the membrane, absorbing some and excreting others back through the membrane to form a primitive cell wall encasing and protecting the delicate membrane itself. This is "metabolism" at its most basic. Of course, this process cannot continue in the meteorite itself but in the parent body at a time when liquid water was present among the matrix. All that now exists in the meteorite samples are the fossil remnants of this process, preserved as the meteorite body dried out.

The same basic structures were found in all three of the meteorites that Pflug studied, and probably constitute significant portions of the organic material in all carbonaceous meteorites. Unfortunately, though, their study does not seem to have progressed since Pflug's papers were first published in the early 1980s.

Nevertheless, the broader issue of microscopic structures was given a new lease on life in the early years of the present century with renewed claims by a number of researchers that structures closely resembling fossilized bacteria were found within carbonaceous meteorites. Once again, the three old amigos – Orgueil, Murchison, and Allende – were the principal culprits.

In 2004, NASA/NSSTC scientist Richard Hoover announced that he had found features closely resembling fossilized cyanobacterial mats in a freshly fractured interior surface of one of the Orgueil stones. These structures were several orders of magnitude larger than the ones found by Pflug over two decades earlier, and also differed from the isolated inclusions of Nagy et al. Some of the structures were as large as 1–10 µm in diameter, with filament lengths of over 150 µm long in certain instances. But the really interesting feature of these structures is their occurrence as complex assemblages. This, as well as their presence relatively deep within the meteorite fragments, speaks against contamination. It goes without saying, of course, that very stringent measures were taken during the preparation procedure to guard against contamination inside the laboratory.

Hoover and colleagues claim to have identified some of the structures as the carbonized and mineralized remains of phototrophic prokaryotes. It is very unlikely, according to Hoover, that mats of these organisms can be explained in terms of terrestrial contamination following the meteorites' fall.

Significantly, very similar structures were also found in the Murchison and Allende meteorites.

It is to be hoped that these results will breathe renewed life into this controversial subject and that the connection between those various structures found by Hoover, Pflug, and Nagy will eventually become clear.

Presumably these microstructures are common to all carbonaceous meteorites, but the high profile played by the Orgueil, Murchison, and Allende falls depends largely on the amount of material deposited by these events. By the standards of carbonaceous chondrites, these were large falls.

The Allende meteorite was the largest of the three, although the amount of carbonaceous material contained within its matrix was very small. It was classified as a Type III carbonaceous chondrite of the CV chondrite class. In this context, "CV" stands, not for *curriculum vitae*, but for "carbonaceous meteorite of Vigarano type." The Vigarano meteorite, the prototype of this class, fell in Italy on January 22, 1910. Meteorites of this type display evidence of having been subjected to significant degrees of heat and pressure while within their parent body(ies) and more closely resemble ordinary stony meteorites than the more primitive and carbon rich carbonaceous chondrites represented by Orgueil and Murchison.

So what can we say at this moment in time (pending further announcements from Hoover et. al.) about the life potential of meteorites?

There is no doubt that certain classes of these objects do harbor organic compounds that are often of considerable complexity. But "organic" does not necessarily imply "biological," and even those compounds found in meteorites that are biologically significant – nucleic acids, for example – need not necessarily be the products of biological processes. Even so, it is interesting to speculate that some or even all of the complex organic compounds forming the chemical basis of life on Earth may originally have come here from outer space.

The ultrafine structures found by Pflug, if they have been interpreted correctly, take us a step further in so far as they show not just the basic life chemistry but also the very basic structure of living cells. Even an extremely rudimentary form of metabolism appears to have been present.

Nevertheless, if the larger and apparently more complex structures found by Hoover and colleagues have been correctly interpreted, the situation goes to a whole new level. If these things really were alive in the meteorites' parent bodies, what was transported to Earth may have been not merely the chemicals of life, nor the structure of living cells, nor even the rudiments of metabolism, but nothing short of full-fledged life itself! This is, of course, a very controversial position to hold, and at this time it can neither be proved nor disproved.

Without coming down on one side or the other, let's consider for a few moments what it would mean if the most elementary forms of life arose, not in a "warm little puddle" on planet Earth, but inside the parent bodies of carbonaceous meteorites.

For starters, what were these parent bodies?

Opinion is divided as to whether carbonaceous chondrites come from C-type asteroids or comets, or from a mixture of both. In reality, the question is no longer so clear cut. In recent years, four apparently normal C-type asteroids have been observed to experience outbursts of cometary activity and have been included in the catalogs of both classes of object. Undoubtedly, these examples represent the mere tip of what is probably a very large iceberg.

Interestingly, a large family of C-type asteroids known as the Themis family is known to host two of these part-time comets. Moreover, its largest member (24 Themis, for which the family is named) is now known to have both ice and complex organic compounds spread throughout its surface material. In all likelihood, most if not all of the members of the Themis family have experienced cometary phases at some time, and it is certainly responsible for a ring of dust – initially found in the early 1980s in data from the infrared satellite IRAS – within the Asteroid Belt. The lion's share of this dust has been released through collisions between the family's member asteroids and smaller meteoritic objects. Probably, these impacts also expose internal stores of ice and thereby trigger cometary outbursts, which add a little more dust to the environment. Gradually, this dust spirals inward toward the Sun. On the way, some of it gets swept up in Earth's atmosphere and filters down to the surface. It is thought that much of the hydrated carbonaceous dust that reaches Earth originated in the Themis family. The spectrum of these asteroids is also a very close match to that of the Murchison meteorite.

Besides Themis and its many siblings, thousands of other C-type asteroids (some of which have also been seen to double as sometime comets) inhabit the Solar System. If these harbor or once harbored living organisms, there must have been a lot of biological activity beyond Earth in the early days of the Solar System. Collisions between asteroids, meteorite impacts on asteroids, and bursts of cometary activity probably all added to the expulsion of dust particles – some of which presumably carried the spore of microorganisms if this scenario is true – into interplanetary space. As cosmic dust floats down onto Earth today, so these particles floated down to the primordial Earth. Carbonaceous chondrite meteorites, early cousins of Allende, Orgueil, and Murchison presumably fell to Earth then as today, except that in very early times, they may have carried not microfossils but viable spore. An interesting thought, is it not?

At this point it may be worth mentioning that the spectra of a Leonid fireball as bright as the full Moon seen in 1999 also provoked a minor controversy. There is no doubt that organic compounds were evident in its spectrum, but an analysis by a scientific team led by Chandra Wickramasinghe went a step further and concluded that the organic signature actually indicated the presence of living microbes; at least, they would have been living until incinerated by the meteor's flight through our atmosphere!

Most scientists do not agree with this interpretation. Organic, yes. Biological? Probably not.

Cook a bacterium or cook some abiotic organic material and you get the spectral features observed. There is no sure way of telling the difference, and the simplest (and therefore most probable) interpretation remains the abiotic one. But, as always, we must remember when considering this matter that the mind is like a parachute; it only works properly when it is open!

By the way, the Leonid fireball did not produce a meteorite. The meteors of this shower move much too fast through the atmosphere to survive. Even if they entered the atmosphere at far lower velocities, it is improbable that they would survive as meteorites. They are simply too fragile and, with a possible rare exception, too small. Yet, their composition is probably not very different from carbonaceous chondrites of the Orgueil type, albeit even more friable in their structure.

The potential biological environment would be even larger if long-period comets turn out to be suitable homes for microorganisms.

Estimates as to how many of these objects orbit our Sun at great distances vary, but even the most conservative places the number in the tens of billions, with a hundred billion or thereabouts being a regularly quoted figure. Some astronomers would increase this even further by one or two powers of ten.

But whatever the actual number, we can be pretty sure that it is but the tip of the iceberg when it comes to the original population of comets that formed with the Solar System. An estimated 95% of the original store of comets escaped the Solar System altogether, thanks to the gravitational perturbations of the giant planets Jupiter and (to a much smaller extent) Saturn, Uranus, and Neptune. The tens or hundreds of billions that remained behind are those that failed to make the great escape. The rest of the "berg" – the greater majority of the Solar System's original store – lies scattered across the galaxy where, presumably, they have been joined by similar escapees from other planetary systems. Do all of these carry organic, even biological, material? We don't know, but the prospect is an interesting thought.

Dr. Waltemath's Many Moons

As we have seen, reliable people at times come up with seemingly incredible tales. The present story is not, however, one of these. The tale we are now about to tell is one in which a highly *unreli-able* person gave voice to an equally unreliable story. We might almost call it lunacy!

This is the story of one Dr. Waltemath of Hamburg and his obsession with moons. Not the one we all know and love, but other moons of our planet – very peculiar terrestrial satellites with some highly unusual properties.

The "Waltemath Moons" first made their appearance in the late 1890s in a series of anti-astronomical-establishment publi-cations with the auspicious title *Astronomical Reports: Organ of the Union for Investigation of the Dark Moon of the Earth*. These "reports" seemed to have been largely devoted to pouring buckets of verbal vitriol over the heads of certain leading astronomers of the day, but they did find room for detailed accounts of Waltemath's own discoveries of a system of small and dark moons orbiting our planet (unconfirmed, of course, by any "establishment" astronomer).

With an occasional exception (about which we shall speak in a moment), these moons were unobservable. This was not because of their small size. It had more to do with their strange composition. You see, unlike all other known material objects, the moons of Dr. Waltemath absorbed *all* incident light instead of reflecting some back into space, as even the blackest "normal" objects do. In a sense, they were a little like the so-called black holes about which so much has been written in recent decades except that, unlike black holes, their strange light-absorbing ability did not relate to their powerful gravitational field. Just as well, really, as the thought of black holes floating around in near-Earth space is not a comfortable one.

Somehow, in spite of the difficulty in discovering something that absorbs light instead of reflecting it, Dr. Waltemath must have made a sufficient number of observations of his principal moon to allow its orbit to be calculated and its dimensions determined. In his own words "Its distance from Earth is 640,000 miles; its diameter is 435 miles. It is faint generally and can only be seen with the aid of a large telescope". [If it absorbed all light, would it not be completely invisible and not just "faint"?].

Earlier, we noted that there was an exception to this moon's non-observability. In the good doctor's own words, the little moon "sometimes ... shines at night like the Sun." As supporting evidence for this assertion, he stated that Lieutenant Greely actually saw this moon when in Greenland in 1881 but mistook it for the Sun itself!

The new moon was also, it seems, accredited with the ability to periodically alter our planet's climate! Noting that "one hundred and six anomalistic rotations of the new satellite are almost exactly equal to the 35-year period in climatic changes established by Professor Bruckner," Waltemath suggested that the satellite might be responsible, although (conveniently) he did not suggest how the moon's climatic influence might be accomplished.

At this point, the reader might be wondering whether the doctor was serious, or whether he had other problems as well, but (strange though it seems) his peculiar ideas met with an equally peculiar response. Some normally conservative astronomical journals, remarkably, afforded him the honor of serious discussion, and one – the prestigious *Astronomische Nachrichten* – even went so far as to publish a series of alleged observations of a solar eclipse by the "moon"!

This alleged event had been predicted by Waltemath for February 4, 1898. By that time, he had already determined that his principal moon was in actual fact one of a whole family of satellites and that eclipses by these would also take place during the preceding two days. It appears that these went unobserved (no surprise here!). The eclipse on Day Three, by contrast, had the effect of "legitimizing" the event in the public mind, and this may have excited the expectations of some to the degree where the event was "observed," whether it actually occurred or not! Be that as it may, the eclipse was allegedly witnessed by a group of twelve people at the Greifswald post office.

Writing of the fiasco, many years later, in the magazine *Sky & Telescope*, Joseph Ashbrook gave a mildly amusing account of what he believed really happened on that day. He drew a mental picture of the "faintly preposterous scene" of an "imposing-looking Prussian civil servant pointing skyward through his office window" while reading "Waltemath's prediction aloud to a knot of respectful subordinates." We could imagine the dutiful employees squinting at the Sun with their unaided eyes in search of the black dot that their superior had insisted was crossing its face. Nobody would have dared opine that the emperor really *was* naked after all!

It is painfully (hah!) easy to "see" dark objects silhouetted against the Sun during a naked-eye glimpse in that direction, especially if one knows what one is *supposed* to be looking for. It is not too surprising that these people (none of whom made a practice of looking at the Sun, we may safely assume) "saw" what they wanted – or more accurately, were told by their superior – to see. Of course, regular solar observers along the path of the supposed eclipse saw absolutely nothing unusual.

Perhaps we should let the final word on this matter go to the judiciously anonymous contributor to *English Mechanic*, July 29, 1898, who recounted observations made on February 5 of a second eclipse, probably by the second of Dr. Waltemath's moons. This eclipse had also been observed, our anonymous correspondent informs his readers, "by three German officers in China," but no further details are given.

Nevertheless, the mystery correspondent predicted yet another eclipse by this second moon "on or about July 30, a few days sooner or later." This one was likely to be very interesting, as our correspondent also informed that the "third moon will pass before the Sun about the same time," implying that the Sun was about to be eclipsed by two moons at the same time! And what an eclipse it would be; the third moon "is larger than the second" and apparently pretty slow as well. The eclipse would last for about an hour and a half. Moreover, *this* was the moon responsible for the Greifswald eclipse of February 4 (or so our correspondent informs us).

Finally, and with (we hope!) his tongue now firmly in his cheek, our correspondent writes that "a second moon ... is no uncommon phenomenon at certain hours of the night." This may well be true, especially if the observer is returning home from a very lengthy and exuberant celebration!

Strike a Light, Another Potassium Flare!

Sometimes in astronomy, as in life in general, not everything is quite the way it seems!

The following story might be described as amusing, embarrassing, or simply as a caution not to ascribe anomalous observations too quickly to unknown phenomena before more mundane solutions have definitively been eliminated.

Between the years 1962 and 1965, astronomers at the Haute Provence Observatory in southern France observed, on three separate occasions, unexpected strong and very brief potassium flares in the spectra of otherwise unexceptional dwarf stars. Strong potassium emission lines would appear without warning, last for a few seconds, and then fade away.

What could cause such an event? And why should this occur in otherwise normal and completely unexceptional stars? Were these flares evidence of some process within apparently ordinary stars that had so far eluded astrophysical modeling?

Before there could be any attempt at revising the accepted stellar models, the phenomenon had to be independently verified, and to this end a group of University of California astronomers attempted to duplicate the observations at California's Lick Observatory. Although the Californian astronomers worked with the same type of spectrographic equipment as their French counterparts, nothing resembling potassium flares were recorded.

Then, one of the astronomers had a bright idea. He struck a match and ... a beautiful potassium flare graced the spectrogram!

It turned out that the night assistant at Haute Provence Observatory on the nights in question was a smoker. The potassium flares were, it seems, produced when he (not the dwarf stars) lit up!

There are three morals to this tale. First, carefully check the possibility that extraneous light might be "polluting" an observation. Second, be cautious about reporting anomalous observations until all alternatives have been eliminated. And third, quit the filthy habit! Not only does smoking rot your lungs and clog up your arteries, it also plays havoc with your astronomy!

Two Meteoric Mysteries

The Tunguska Enigma

Few natural events of the last century have gripped the imagination as much as the mighty blast that occurred over the lower Stony Tunguska River in Siberia on the morning of June 30, 1908. It has become the stuff of great speculation – some restrained, some totally wild – as well as a good deal of serious research conducted in what must be one of the most mosquito-ridden regions of this planet.

Because of the remoteness of the region, news of the event only trickled out over the course of years, but when the scraps of information were eventually pieced together, the following story emerged.

Around 2 min past seven o'clock, local time, on the morning of June 30, (June 17 on the Julian calendar used locally at the time), a brilliant object almost as bright as the Sun, albeit of a bluish color, moved across the skies and exploded some 3–6 miles (5–10 km) above the ground in what was, in that pre-atomic age, one of the largest explosions ever witnessed by humans. One eyewitness told how the "sky split in two and fire appeared high and wide over the forest [until] the entire northern side [of the sky] was covered with fire." This same witness recounted being thrown to the ground by a hot wind, before briefly losing consciousness.

The sound of the explosion was said to have resembled "rocks ... falling and canons firing" or "some kind of artillery barrage."

Estimates of the power of the explosion differ. A yield of somewhere between 10 and 15 megatons, or around 1,000 times that of the atomic bombs dropped on Hiroshima and Nagasaki at the close of World War II, is often quoted, although some estimates have gone as high as 30 megatons. More recent research has, however, been more modest, with estimates of "only" 3–5 megatons. But whatever the true figure, the power of the explosion was very evident. Some 80 million trees were flattened over an area of 830 square miles (2,150² km) and an estimated 1,500 reindeer were killed. An earthquake, retrospectively estimated at about five on the Richter scale (which had not been developed at the time of the event), spread out from the center of the blast. The shock wave from the explosion registered on micrographs in England and one day later in Potsdam, after having navigated the entire sphere of Earth.

For several weeks after the blast, night skies across Europe and western Siberia were unusually bright. It is said that newspapers could be read out of doors as far south as the Caucasus by the light from the sky alone. The atmosphere also became unusually hazy, with a noticeable decrease in atmospheric transparency as far away as California from mid July until late August 1908.

Of course, all of these anomalous observations remained apparently unconnected until the story was finally put together years later. In one sense this was fortunate. Had the object arrived a little less than 5 h later, it would have scored a bull's-eye on the city of St. Petersburg, the capital of what was then Imperial Russia.

Nevertheless, the strange stories of lights and explosions in that remote region were not entirely forgotten. Wars and revolutions did not totally erase them from the collective Russian memory, and in 1921 Russian mineralogist Leonid Kulik visited the region as part of a survey for the new Soviet Academy of Sciences. He saw enough to convince him that a large meteorite had been responsible for the dramatic events of 1908 and managed to convince the Soviet government to fund a more comprehensive expedition in 1927. This was followed by three more

This scene of devastation awaited the Kulik expedition, some 13 years after the Tunguska event. L. Kulik, Soviet Academy of Sciences 1927.

during the next decade, culminating in aerial photography of the region in 1938.

What Kulik expected to find, however, proved elusive. At the apparent point of the explosion, there was almost total devastation of the forest, but no obvious meteorite crater. Some small depressions initially thought to be shallow craters were found, but these turned out to be simple formations common throughout the region and quite unconnected with the meteorite.

The Kulik photographs did, however, turn up a butterfly-shaped pattern in the devastation, indicative of an aerial blast. This would also explain the absence of a crater. But what type of giant meteorite blows up before reaching the ground?

It is this latter fact that has led to all sorts of ideas about what the Tunguska object might have been. Proposals have ranged over just about everything from a mini black hole to an atomic reactor accident on board a flying saucer!

At a more serious level, it was early noted that the date of the event was close to the time when Earth passed close to the then orbit of a short-period comet known as Pons–Winnecke (because this comet has been tossed to and fro by the gravity of Jupiter, it currently follows quite a different path – but that is another story!). On some years, a meteor shower was noted as we encountered dust left behind by Pons–Winnecke, and the suggestion was made that a piece of the comet's nucleus might have come loose and traveled around its orbit together with the meteors, finally running into Earth in 1908. Unfortunately for this suggestion, a fragment of P–W would have approached from quite a different direction from that of the Tunguska object.

Nevertheless, the comet hypothesis was persistent. It seemed to explain the aerial explosion (a low-density comet would be more likely to explode mid-air than a denser meteorite). It also might explain the bright skies following the event. Maybe this was due to the comet's tail dispersing in Earth's atmosphere.

The time of year (late June) also appeared suggestive to some astronomers in more recent decades. Remember the Canterbury Swarm and the great lunar flare? The co-incidence with the Beta Taurid meteor stream and its Comet Encke association seemed too much of a coincidence for some. Perhaps the Tunguska object was a piece of a comet after all – but a fragment of Encke rather than of Pons–Winnecke.

Apart from the coincidence of dates, however, there is little to identify the Tunguska object as a fragment of Encke, and a re-evaluation of the evidence by Z. Sekanina in the 1980s, taking into account evidence from the blast pattern plus old eyewitness reports, enabled an approximate orbit to be derived for the Tunguska object, which looks very similar to an ordinary Apollo asteroid. From the degree of penetration into the atmosphere, Sekanina also estimated the density of the object to be similar to an ordinary stony asteroid. A comet, according to his computations, would have exploded at a higher altitude, while an iron-rich asteroid would have penetrated much deeper, even smashing into Earth's surface in a crater-causing explosion.

The mystery of an aerial explosion is easily solved. The atmosphere exerts a tremendous pressure on objects moving rapidly through it, and unless they are strong and dense, exploding is the natural thing to do. Many "mini-Tunguskas" are seen, some of them not all that mini in actual fact. Kiloton blasts high in the atmosphere are not nearly as rare as might be thought!

Nevertheless, despite this quite logical explanation as to why the object did not produce a crater, new research has raised the possibility that it might have left one after all!

Earth approaching asteroid. A Tunguska-sized stony asteroid approaches Earth, as depicted in this painting by Dr. William Hartmann. ©William K. Hartmann, January 1994.

Map of the Tunguska impact region.

About 5 miles northwest from the center of the explosion, there lies a small freshwater lake known as Lake Cheko. Some suspicions were long ago aroused about a possible association between this lake and the explosion, but a 1961 investigation found the depth of silt at the lake floor too deep for it to have been formed as recently as 1908, and an estimated age closer to 5,000 years was given.

On the other hand, there is no definitive evidence that the lake was there prior to 1908, and tales circulate in the area specifically saying that it was not. Moreover, an expedition in the summer of 2007 by L. Gasperini and colleagues from the University of Bologna raised a number of doubts about the lake's relative antiquity. The sediments are certainly quite deep, but a sounding of the lake bed revealed a "hairy" look about them, which could be interpreted as being due to branches of fallen trees. If the "sediments" are the remains of a forest, their depth might not be a reflection of their age.

Moreover, the shape of the lake is unusual and appears to set it apart from others in the region. Thanks to many years of sedimentation, most of the nearby small lakes are shallow and (for want of a better expression) more or less saucer-shaped. But Cheko is more conical, falling away quite quickly to a considerable depth. This has been interpreted as a sign of the lake's youth. Lake Cheko has not had time to silt up and turn into a shallow saucer.

Maybe it's a coincidence or maybe it is telling us something important, but the long axis of the lake also points in the direction of the trajectory of the Tunguska object. And, if all of this is not enough to at least arouse suspicion, magnetic readings indicate that there might be a chunk of rock over a yard in diameter buried below its deepest point. Could this be the last remaining fragment of the Tunguska meteorite?

Still, there are skeptics. Garath Collins and Phil Bland of Imperial College in London point out that many of the trees surrounding the lake are clearly older than a century and must therefore have stood from a time before the explosion. If Cheko really is the elusive Tunguska crater, it is not easy to understand how they avoided being blown flat.

So, at the moment, the case is far from closed. As these words are being written, though, preparations are being made for a further

Bologna expedition that will concentrate on the lake and attempt to decipher whether a mass really does exist under its deepest point and, if so, how large it really is and what it might be.

Perhaps the final answer to the Tunguska mystery is not far off. But then again ...

Tunguska impact: moment of explosion from Vanavara Trading Station. © William K. Hartmann, August 1996.

Tunguska: A minute after explosion. © William K. Hartmann, August 1997.

Tunguska fireball: Seen from Kirensk. © William K. Hartmann, August 1995.

A distant view of Tunguska. "Where the body disappeared behind the horizon, a pillar of dark smoke rose up". © William K. Hartmann, May 2001.

The Fireworks of St. Cyril

Meteor showers are traditionally named after the constellation from which their members appear to radiate. Thus, as previously noted, we have the Leonids from Leo, the Orionids from Orion, and the Quadrantids from the old constellation of Quadrans Muralis (the Mural Quandrant), long since absorbed into Bootes. A few are alternatively called by the name of the comet that spawned them. The Andromedids, for example, are also known as the Bielids, in honor of Biela's Comet, whose debris formed the stream.

But what are we to make of the Cyrillids?

The name is in honor of St. Cyril of Alexandria, on whose feast day (February 9) they were seen.

Actually, the Cyrillids were not really a meteor shower at all, which is the reason why we are interested in them here. The story of the Cyrillids is (well) quite weird!

It was on the evening of the Feast of St. Cyril, February 9, 1913, when the strange event occurred, an event probably not as well known nor as (in-?)famous as Tunguska, but in its own way no less peculiar. At around 9.15 p.m. local time, on that winter's evening, many residents of Canada and parts of the United States witnessed a parade of fireballs flying in a narrow corridor from the northwest toward the southeast. Some witnesses counted 15 separate objects, but these must have been just the brightest of the parade, as others claimed that "thousands" of meteors followed one another across the sky (though this number is probably an exaggeration). Unlike a genuine meteor shower, the fiery objects did not appear to radiate from a small area of sky; the "radiant," as explained in Chap. 4, is a perspective effect resulting from the near-parallel orbits of meteors belonging to a true shower. By contrast, the Cyrillids appeared to be following one another in a narrow procession.

The display, even if not a genuine shower, must have been truly spectacular. According to a *Popular Astronomy* article by W. H. Pickering, published 9 years after the event, most of the fireballs were of a yellow or reddish color, the majority resembling bright stars or even Venus in magnitude. However, the leading objects were larger, appearing about the size of the Moon and shining with more of a violet hue. According to this article, the fireballs were "arranged at first in four or five separate groups" and were accompanied along part of their rout by "innumerable finer particles that were swept off them in their rush through our atmosphere." This last remark probably accounts for the claims that "thousands" of meteors were seen that evening.

Certain of the objects were said to have sported long tails. According to some reports, the display was accompanied by a rumbling sound that caused buildings along the path of the meteors to vibrate.

From talks with eyewitnesses to the phenomenon, C. A. Chant, who published a paper on the event in the year of its occurrence, concluded that the procession of fireballs lasted for about 3.3 min. This is quite long for a meteoric event, although much shorter than a true meteor shower. These can go on for weeks!

Other fireball processions have been noted from time to time, and, indeed, they have increased since the beginning of the space age. Are our rockets and artificial satellites drawing strange meteors toward Earth?

Of course not! Artificial satellites *become* strange-looking "meteors" as they re-enter our atmosphere. Indeed, without shutting the door on natural fireball processions, most reports of streams of multiple, slow fireballs reported nowadays turn out to be caused by the atmospheric re-entry of space debris.

Of course, there were no artificial satellites in 1913, but might there have been a *natural* one whose steadily decaying orbit finally brought to a fiery end in our atmosphere?

That was the conclusion reached by Chant and other early investigators of the event. Since the space age, and the many examples of decaying artificial satellites in the atmosphere, the explanation appears even more enticing.

Chant's conclusion does, however, have its skeptics. One scientist who poured cold water, not just on the satellite hypothesis but even on the Cyrillids themselves, was C. C. Wylie, who attempted to demolish the entire story in an article in *Science* in 1953. Wylie was concerned that the display was not witnessed from an even wider area and concluded that it was probably no more than a quite normal display of meteors, accompanied by an equally "normal" fireball that disintegrated into a procession of sparks during its atmospheric flight.

This deconstruction of the account hardly does credit to the eyewitness accounts collected by Chant, nor to a couple of further observations made at sea and collected by Denning subsequent to Chant's publication of his findings. The Denning records add further support to the strangeness of the event and were used by Pickering to further refine Chant's derivation of the fireballs' path.

Wylie was correct, however, in criticizing some later and more sensationalist stories. Some popular accounts went as far as claiming that, had the fireballs not passed over the sea, they would have caused vast conflagrations in and around New York City. This is, of course, pure sensationalist fiction.

The decaying satellite hypothesis does not, let it be stressed, raise the ghost of Dr. Waltemath. It does not imply that our planet had a second moon from its formation until as recently as 1913. We know from celestial dynamics that Earth and other planets can capture asteroids into temporary satellite orbits, and it seems quite possible that some of these might decay until the temporary satellite eventually enters Earth's atmosphere and burns up. If the body was quite fragile, as many asteroids are, it might be broken up by tidal effects even before encountering the atmosphere. If that were the case, a string of objects would enter the atmosphere, following one another in procession along the same orbit. This might explain why the Cyrillid event was of relatively long duration – longer, perhaps, than one might reasonably expect for the alternative possibility of a single object breaking apart *following* its atmospheric entry.

The possibility of tiny moonlets in Earth orbit was raised earlier in relation to anomalous observations of small black spots seen transiting the Moon's face. Perhaps, in the fireworks of St. Cyril's Day in 1913, we have further evidence for the presence of objects of this nature.

Serendipitous Discoveries

In a sense, lots of discoveries are serendipitous. While searching for something else, a new object or phenomenon is found that may even end up being of greater importance than the thing being sought in the first place.

In astronomy, comets are notorious for being discovered by someone engaged in a regular observing program. They have been found in the fields of deep-sky objects, other unrelated comets, variable stars, and planets. Not long ago, an observer found one while looking at Saturn. Others have been found by people testing telescopes and eyepieces.

The classic example of the latter happened to Greenwich Observatory astronomer and double star expert M. P. Candy on the evening of the day after Christmas, 1960. Candy was at the observatory early in the evening measuring some double stars, but the night started to cloud over, and he figured that staying on would be useless.

Soon after arriving back home, he noticed that there was a partial clearing in the north and, while not enough for serious astronomy, he decided to make use of the clear patch by trying out a small portable telescope he had not yet tested on the sky. Setting up the small 'scope inside his home, he focused through an open window on a small section of clear sky not far from the north celestial pole and saw not just a field of stars but a previously unknown and relatively bright telescopic comet! Thereafter named for Candy, the comet became jocularly known at the time as the Christmas Candy comet!

This pales in comparison, however, with the following two incidents.

On November 15, 1890, T. Zona of the Palermo Observatory in Sicily discovered a relatively bright telescopic comet, and a telegram announcing the discovery was quickly dispatched to observatories around the world. On the night of November 16–17, a copy of the telegram arrived at Vienna Observatory, where Rudolf Spitaler was on duty and attempted to observe the object through the observatory's 27-in. (68.6 cm) telescope. Swinging the telescope to the approximate position given by Zona, Spitaler quickly located a comet, but it was a puny little thing – nowhere near as bright as Zona's telegram had led him to believe.

Nevertheless, Spitaler carefully measured the object's motion over the course of 30 min and found clear movement, though not in the direction that he had expected. The comet's direction of motion was not consistent with it having been in the position given by Zona on the previous evening.

Suspecting that something was amiss, Spitaler swept the telescope over a larger area of sky and quickly came upon another nice bright comet. *That* was Zona's. The little one (subsequently named for Spitaler) was a totally unassociated object that just happened to be passing through the same region of the heavens.

One astronomer remarked that the Spitaler incident was such a remote co-incidence that nothing like it was ever likely to happen again. How wrong he was!

On the morning of November 17, 1895, Charles Dillon Perrine of Lick Observatory found a comet just a little too faint for naked-eye visibility. In the following weeks it moved south and brightened but became overwhelmed by morning twilight after December 11. It made a brief and quite brilliant appearance in the southern hemisphere for a few days before Christmas, when it was visible with the naked eye and sported a bright tail, before vanishing again into the twilight. There it remained until February 1896, when it reappeared in the northern hemisphere morning skies as a telescopic object.

This is where the story gets a little weird!

One of the first astronomers to recover the comet was Dr. E. A. Lamp at Kiel Observatory in Germany, who sighted it on February 13 and sent out a telegram announcing the observation. Somehow, as the telegram was distributed to American observatories, the message became garbled and an erroneous position was given.

In the meantime, Perrine himself had also recovered the comet (actually a few days prior to Lamp) and, when he saw the (erroneous) position given in the telegram, concluded that the comet seen by Lamp was not the one he had discovered the previous November. Suspecting that Lamp had in reality spotted a new comet, Perrine swung the Lick telescope to the position given on the garbled telegram and (unbelievably!) *found* a new comet at that location, which he naturally assumed to be the one that Lamp had discovered!

But more confusion was to follow.

Thinking that he had confirmed "Lamp's Comet," Perrine notified Boston (then the clearing house for astronomical announcements), which telegraphed the message to the rest of the astronomical world. Lamp, on receiving Perrine's "confirmation," wondered what was happening. He had, after all, located Perrine's earlier comet at exactly the spot where it should have been! Suspecting, as

Perrine had earlier done, that a new comet had been found (but this time by Perrine!), Lamp swept the area and, sure enough, located the comet that Perrine had observed and misidentified as Lamp's.

Eventually the mess was sorted out and both comets verified. They were totally unrelated objects, which just happened to be passing like ships in the night. Around February 16, they were located just 4° apart in the morning sky.

By the way, when it came to officially naming the 1896 object, both Perrine and Lamp were recognized, and the comet is now known as Comet Perrine-Lamp.

Only slightly less odd were the circumstances surrounding the discovery of comets Levy and Shoemaker-Holt in 1988.

On March 19, 1988, amateur astronomer David Levy came across a small and generally unremarkable comet while sweeping the morning skies. Preliminary orbital computations showed that it would pass a little outside of Earth's orbit and never become a bright object.

On the morning of May 13, Eugene and Carolyn Shoemaker, together with Henry Holt, finished their minor planet search program at Palomar Mountain Observatory somewhat early and, as there was still some darkness left, decided to try for an image of Levy's Comet. The telescope was duly set and an exposure made that nicely captured the comet near center frame.

However, something was clearly wrong. The stars in the field could not be identified!

Checking through their procedure, the Shoemakers found that an error had been made in the positioning of the telescope, and the field photographed was actually some 17° from the true position of Levy's Comet. The comet in the photograph had to be a different one!

Unlike the above two examples, comets Levy and Shoemaker-Holt (as the new object was subsequently named) were not unrelated. They were following one another along the same orbit, except that their dates of perihelion (closest approach to the Sun) were separated by 76 days.

The two objects are, apparently, major fragments of a single object that split in two sometime between its last journey through the inner Solar System (thousands of years ago) and 1988. This relationship between the two comets does not, however, make the circumstances of the second object's discovery any less remarkable.

Our final example of a weirdly serendipitous comet discovery is a little known one involving Halley's Comet at its most recent return. The details were supplied by Maik Meyer, who got most of the information directly from the astronomer involved, so there are very few degrees of separation in this account!

The story involves German astronomer Ulrich Thiele who, on the night of October 9, 1985, was on duty at the Calar Alto Observatory in Spain. At the time, the observatory was engaged in a program of regular observations of the approaching Halley's Comet.

The night in question was not good. Cloud covered the sky for much of the time, and sometimes the observatory was shrouded in fog. Around five o'clock in the morning, Thiele decided to retire to bed and turned off everything in the building. Stepping outside, however, he noticed that the sky had unexpectedly cleared and, hurrying back inside, put two photographic plates in the telescope and turned on the power. He aimed the instrument for the coordinates of Halley and began the exposure. In his hurry, however, he omitted precessing the coordinates to the actual date, which meant that the telescope was still centered on the position for the standard equinox of 1950. Because of the wide (5.5°) field of the Schmidt telescope being used, this was not a serious oversight, but it did mean that the comet was off-center in the photographs.

Nevertheless, a problem arose because an emission nebula also lay near the plate's center. The comet was not the only nebulous object in the field and, being off center, had to be sought out in the wide field image.

At the time he conducted this search, Thiele had not become aware of his error in using the wrong coordinates to align the telescope and had to search the entire plate to positively identify Halley. It was during this search that he came across a previously unknown comet faintly visible close to the edge of the photographed field! Meyer notes that three coincidences joined hands to bring about the discovery of Comet Thiele.

First, had the telescope been aligned to the correct coordinates, the new comet would have fallen outside of the photographed field.

Secondly, because Halley was not easily visible within the emission nebula, a search of the plate was necessary to identify it. Had no nebula been present, that comet would have been obvious and a further search of the field not required.

And thirdly, just after the two plates had been exposed, the skies again clouded over and fog once more enshrouded the observatory!

Comets are not the only objects that have been serendipitously discovered. A very interesting example involves the white dwarf companion of Sirius – Sirius B or "the Pup" as it is popularly known (see Chap. 5). This object is not especially faint. It would be quite easily seen in small telescopes were it not for the close proximity of the overwhelmingly bright Sirius A. Had it been located somewhat further from Sirius A, the Pup would no doubt have been found far earlier. In actual fact, this tiny but ultra-dense star was discovered completely by accident.

The story runs like this: On the night of January 31, 1862, Alvan G. Clark, Jr. (son of the great telescope maker), was using Sirius as a test object for a newly completed 18.5-in. telescopic objective at his father's optical shop in Cambridgeport, Massachusetts. It was then that he noticed a small and (by comparison) faint point of light very close to the star.

Now, faint pinpricks of light close to bright stars not infrequently turn out to be optical ghosts. Indeed, an optical ghost near Procyon later fooled some highly experienced astronomers into thinking that they had found a small companion of that star as well (in truth, Procyon really *does* have a companion, but this is a far more difficult object to observe than Sirius B and remained unobserved until 1896, although its presence was inferred as early as 1844 through its gravitational perturbations on Procyon A).

Clark's pinpoint of light proved not to be a ghost, however, but a real star, and an important one at that – a white dwarf almost on our back doorstep, cosmically speaking. Clark's must be listed as one of the more important serendipitous discoveries.

What Was the "Christmas Star"?

Although this may not be a strictly astronomical story, the "Christmas Star" or "Star of Bethlehem" or "Star of the Magi" is so well known and of such long history that the subject is bound to be raised in general astronomical discussions. In the Western world just about everyone – believer or unbeliever – has heard some version of the story. The star is represented on Christmas trees (vying with an angel as the top-most decoration), in Christmas cards, nativity scenes and plays, and so forth. At least two species of flower are popularly known as "Star of Bethlehem" lilies.

Unfortunately, much of the popular, sentimental lore surrounding the Star bears little resemblance to the original account as it appears in the *Bible*. At least two Christmas songs have the shepherds (not the Magi) seeing the Star, one plaintive and beautiful song has Mary and Joseph following it, while another venerable carol says that the star was seen both day and night. For someone seeking an astronomical explanation, these traditions can become veritable red herrings.

Over the years, just about every conceivable astronomical phenomenon has been put forward as an explanation, as well as skeptical suggestions that the entire story is nothing more than myth. Some of these suggested explanations have run afoul of the red herrings mentioned in the previous paragraph! Any serious attempt at deciphering the nature of the star must begin not with the apocryphal stories that have accrued over the ages but with the original biblical account itself.

Somebody knowing only the "Christmas card" version would no doubt be shocked to find that the star is hardly spoken about in the bible at all. Its only mention is in one of the gospel nativity accounts, as we shall see shortly.

Three reports exist of the nativity of Jesus, in what are known as the synoptic gospels (the accounts given by Matthew, Mark, and Luke). The gospel according to John is more in the nature of a meditation on the theological/philosophical significance of the nativity, rather than a straightforward historical account.

Just looking at the synoptic gospels, it is somewhat surprising (in view of the relatively high profile given to the star in more modern Christmas symbolism) that there is not the slightest hint of its appearance in two of the synoptic accounts. The only evidence for the star in the entire bible is a brief and matter-of-fact mention in the gospel according to Matthew.

In summary, Matthew's account says simply that a group of "wise men" – magi or astrologers – (traditionally three, but Matthew nowhere gives the number) from "the east" arrived in Jerusalem in search of the newly born "king of the Jews." Their country of origin is not stated, although it is widely thought that they were from Persia. According to one little known story, one of the magi was an Indian pundit by the name of Vishwamitra, who upon returning to his own land gathered a group of followers that later formed the nucleus of a secret organization known as the Secret Sanyasi Mission, said by some to persist to this very day!

Wherever they hailed from, the magi first went to Herod, Rome's puppet king of Palestine, no doubt expecting to find the new prince in the ruling king's palace. They explained their search by telling Herod that they had seen his (the prince's) "star in the east" or "rising" (as newer translations prefer). Herod then inquired as to when the star first appeared. (Apparently, neither he nor his advisers knew anything about it until the magi told him.)

After leaving Herod, the magi again saw the star that went before them until it stood over the place where "the child" was. Incidentally, it is only after leaving Herod that the magi could be said to have "followed" the star. If they originally came from "the east" and first saw the star in the eastern sky, they must have had their backs to it for most of the journey.

That is the extent of Matthew's account. Nothing is said specifically about the star, and, indeed, the impression is given that Matthew is not particularly interested in it. His chief interest concerned the magi and their journey in search of the baby Jesus.

Before looking at what the star might have been, some general comments about Matthew's account are needed to place it in a proper perspective. This "proper perspective" is, as we shall soon see, quite different from the one so often assumed.

For a start, there is not the slightest hint in Matthew that the star was anything miraculous or even conspicuous. He does not even say that it was some new object appearing in the sky. Nothing in his very brief mention precludes the possibility that it was one of the firmament's regular denizens that had for some reason assumed special significance to the magi (we will return to this point soon). There is no suggestion that anyone other than the magi knew of it; specifically (as remarked above) neither Herod nor his advisers appear to have known of it prior to the magi's arrival. Yet, surely something conspicuous and/or unusual in the skies would have aroused a good deal of attention.

Moreover, contrary to what is often assumed, Matthew does not say that the star *foretold* the birth of Jesus. The magi saw it as a sign, but Matthew only reports this without comment. He does not say that *he* regarded it as a sign nor does he anywhere indicate that his readers should think of it in these terms.

In fact, coming from a Jewish background, Matthew is not likely to have encouraged astrological ideas, even those spawned by a star announcing the birth of the Jewish Messiah. The Jewish scriptures – the Old Testament of the Christian bible – condemned astrology. At the time of Matthew's writing (several years after the events described), these Jewish scriptures were the only sacred writings possessed by the early Christians, and it is hardly to be thought that he would introduce something as dubious as an astrological idea into his account of the nativity. So why did he mention the star at all?

Perhaps his mention of the magi (and, only incidentally, of the star) was directed at his Jewish audience in a way that was both critical and ironic. Scholars agree that Matthew's gospel was primarily written for Christian Jews, and the story of the magi is therefore aimed specifically at them. The irony of Matthew's account is that the Jews – the people privileged with the very oracles of God – by and large rejected Jesus as the prophesied king, while this small band of "heathen" were the first to recognize him. And if that was not enough, they came to this recognition through the practice of astrology – something forbidden to the Jews! It is difficult for us to fully appreciate what a stinging indictment on the closed mindedness of the contemporary Jewish religious authorities this must have been, nor is it easy to gauge its impact on Matthew's first readers. These early Jewish Christians must have gasped at Matthew's nativity account.

This certainly shows the story of the Christmas Star in a different light from that of the rather sentimentalized nativity scenes, but it seems to better fit the original Matthew narrative.

With the above in mind, a number of candidates for the star can be eliminated.

First, we can reject bright and conspicuous transient events. This most probably eliminates bright nova/supernova, comets, and processions of meteors, all of which have been suggested at one time or another.

Secondly, from the above interpretation of Matthew's account we can conclude that explanations reducing the story to myth rest on a misunderstanding of his purpose. The explanation favored by Isaac Asimov, for instance, rejects any historical or astronomical foundation for the story. Asimov suggests that the story of a star announcing the birth of Jesus was a pure fabrication by later writers inspired by their belief in his special importance. This fits with the portents said to have announced the birth of various kings and other great people and may indeed be relevant to the exaggerated apocryphal accounts of the star that appeared long after Matthew wrote (and which bear scant resemblance to his original record). But if the Matthew account itself was not a "birth portent story" – if the apostle had a very different reason for its telling – the "myth" explanation falls apart.

So, if the star was neither a spectacular transient, nor a myth, what could it have been?

One popular theory holds that it was a close planetary conjunction. Despite what our calendars say, scholars agree that Jesus was really born several years earlier than the year zero. He was probably born about 6 BC, and there were indeed some interesting planetary conjunctions around that time. One series of conjunctions was especially interesting, as we shall see in a little while.

However, as linguist Richard Coates points out, the word which Matthew used and which is translated as "star" in English means … exactly that. Just a star, without qualification. A single shining point of light in the sky. The word would not have been used to describe an astronomical event such as a grouping of planets.

What has been said thus far might appear to eliminate everything – including the hypothesis that the star didn't exist! Yet, could it be that the most obvious candidate has been staring us in the face all along? Could it be that the star was exactly that – one of the ordinary fixed stars?

This is the position taken by Coates, on linguistic rather than astronomical grounds. In a talk delivered to the Language Society at the University of Sussex in February 1987 and published in a more extended form in *Astronomy & Geophysics* over 20 years later (October 2008), Coates draws attention to an almost-forgotten pre-Ptolemaic Arabian tradition of naming certain stars *al-sa'd* or (roughly translated) "lucky star." Of particular relevance to the current context is *al-sa'd al-malik* or, alternatively, *al-sa'd al-mulk*, roughly translated as either "the lucky star of the king" or "the lucky star of the kingdom." This is the name used rarely in English as Sadalmelik or Sadalmelek, more formally and popularly known as the star Alpha Aquarii.

Coates thinks it very likely that astrologers of Matthew's time would have referred to stars bearing names of this type. Quite possibly, when the magi referred to "his star" (the star of the new king) they were referring to a star known to them as "the king's star." If they shared the same Arabian tradition mentioned above, that star would very likely have been Alpha Aquarii.

Alpha Aquarii is not an especially bright star as seen with the naked eye. In fact, although designated "Alpha," it is in actual fact very slightly fainter than Beta Aquarii, although the difference is hardly apparent visually. Accurate photometric measurements of Alpha place it at apparent magnitude 2.95 – a "third magnitude star" in less precise terminology.

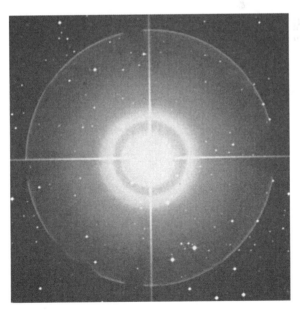

Is this the star of Bethlehem? Sadalmelik as seen by the Hubble Telescope. STScI/SERC.

Nevertheless, we must not be fooled by its relative dimness. In reality, this star is a yellow supergiant having a diameter about 60 times that of our Sun and a luminosity some 3,000 times as great. Only a distance of approximately 800 light years dims its splendor to our eyes. The temperature of its photosphere ("surface") is around 6,000 K, not too different from the Sun's; which is no surprise, seeing that the Sun is also a yellow star of similar spectral type.

Sadalmelik is really a double star. The companion, however, shines only feebly at 12.2 magnitude, less than 1/5,000 the brightness of its supergiant companion.

Identifying the star with Alpha Aquarii, or any other fixed star, begs one huge question however. A fixed star first appears "in the east" at the same time every year. So why did the magi see special significance in its rising *that* year?

It is here, according Coates, that the position of the planets come into their own. A significant alignment between Alpha Aquarii and one of the planets, or a conjunction or planetary grouping at a time deemed auspicious by the magi, would have given the first appearance of Alpha Aquarii a special significance that year.

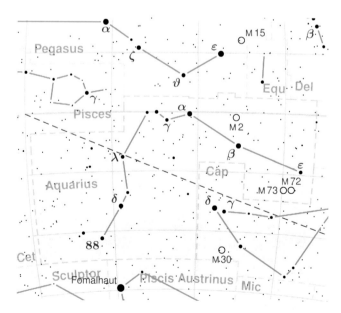

The constellation Aquarius, showing Sadalmelik. Torsten Bronger, source Freebase.

Alpha Aquarii, it should be noted, did not in itself represent any specific king, but if it became aligned with a planet (which, in the astrological lore of the magi represented, for instance, Palestine), astrological logic would give it significance (that year) for the king of the Jews.

In his book *The Bible as History*, Werner Keller recalls a suggestion by Kepler about a conjunction of the planets Jupiter and Saturn in the constellation of Pisces that took place in the year 7 BC. Upon seeing a similar conjunction shortly before Christmas in 1603, Kepler recalled a statement that he once read in the writings of the rabbinic writer Abarbanel referring to a Jewish astrological belief that the Messiah would appear when there was a conjunction of Saturn and Jupiter in Pisces. Wondering if such a conjunction had indeed occurred around the time of Jesus' birth, Kepler calculated the relative positions of these two planets back to this time and found that not one but *three* occurred in 7 BC, all of them in Pisces!

A tradition associating Pisces with Israel, or with the region surrounding it, was apparently well known among eastern astrologers of the time. According to the Chaldeans, the constellation represented the Mediterranean lands (of which Palestine was one), and in Jewish astrological thought (which did exist, despite the scriptural condemnation!) Pisces was the sign of Israel and the Messiah. The constellation stood at the end of the Sun's old course and at the beginning of a new one, so it is not surprising that it should have come to be associated with change and new beginnings.

The two planets were also afforded significance by astrologers. Jupiter was always thought to be a lucky star, and, according to one old Jewish tradition, Saturn represented the protector of Israel. Tacitus even claimed that Saturn represented the God of the Jews.

Such a triple coming together of these two planets in a constellation pregnant with significance for astrologers of the region must have been seen as auspicious.

The sequence of events proceeded as follows.

On May 29, 7 BC the first conjunction occurred. The planets then parted company for a while, only to pair up again on October 3; the Jewish Day of Atonement. The third and final pairing took place on December 4.

Now, following Kepler, Keller and a number of other writers suggest that this triple conjunction *was* the star; however, this hypothesis runs into the problem mentioned above, viz., that Matthew would then be using the word equivalent to "star" in an unconventional way.

However, a little more than 2 months after the final conjunction of the series, around the middle of February in 6 BC, Alpha Aquarii, the "King's Star," again became visible in the morning sky. Following the events of the previous year, would it not be surprising if its rising was given extra significance at that time?

One mystery, however, remains. If the star really was Alpha Aquarii or one of the other fixed stars, how did it lead the magi to Bethlehem and then stand still over the place where Jesus was?

Coates suggests that by the time the Magi reached Jerusalem, the star which had been just showing up in the dawn when they left home would be culminating in the evening sky. From the latitude of Jerusalem, Alpha Aquarii culminates in the southwestern sky. Because Bethlehem is slightly west of south from Jerusalem and about 5 miles, or 8 km, distant, the diurnal movement of the star would have taken it more or less in that direction as the magi set out on their journey from the city. Assuming that they covered the distance in 1 or 2 h, Alpha Aquarii would have "gone before them" until it culminated in the south. At that time, it would briefly appear to pause as its upward movement ceased and its slow decline into the west began. Beneath the spot where it paused, the new king lay!

Well, then, has the mystery of the Christmas Star at last been solved?

Will this finally put an end to debate about the nature of the star?

Probably not. But the good news is that the star is still shining up there for all to see, an ever present reminder of that long-ago event that changed the world.

Project 24
Seeing the Star of Bethlehem
on Christmas Eve

Although scholars agree that Jesus was not born on December 25, the fact remains that a large section of humanity celebrates his nativity on this date, and the Star of Bethlehem has become intimately bound with these festivities. Stars are placed on Christmas trees and drawn on greeting cards. Children love seeing Christmas decorations, but what if their parents could show them the *real* Star of Bethlehem on the night before Christmas? Now that would be something different! If Alpha Aquarii truly is the one the Magi followed, the good news is that it is an early evening object on Christmas Eve, shining there for all to see.

After twilight fades on Christmas Eve night, look toward the western horizon for the constellation of Aquarius. Because Alpha lies essentially on the celestial equator, it will appear a little south of due west for northern hemisphere observers and a little north for southern. It is easy to find once the Water Jug – otherwise known as the steering wheel asterism – is located. This asterism is very distinctive, comprised of four stars marking out a pattern very like the Mercedes insignia. Unfortunately, the stars are quite faint, and although it is easy to see with the naked eye in dark skies, it may require some help from the inner city.

Directly west (beneath) this asterism and about 5° distant from it is a somewhat brighter star (though still not "bright" according to most people's understanding of the term). This is Alpha Aquarii, Sadalmelik, arguably the Star of Bethlehem.

If it looks a little faint with the naked eye, a small telescope will quickly rectify that. Notice that the star has a yellowish tinge (more perceptible through a telescope, but possible to detect by eye alone under good conditions). This is not surprising, as Alpha Aquarii has a spectrum very similar to our Sun, although, as we have seen, it is an otherwise very different type of star.

It is fitting to end this excursion through the byways of astronomical happenings with the story of the Christmas Star. Our ramblings have taken us from the ridiculous in the form of Waltemath's supposed dark moons and Abba 1, through genuine but misinterpreted observations, honest mistakes, genuine mysteries, and serendipitous discoveries possessing an importance unknown to those who made them. And so we end with the sublime, an ancient story known to a great part of the world's population but still the subject of much speculation.

As we said at the outset, it is unlikely that anyone reading these words will now know a great deal more about astrophysics or the deep speculations of cosmologists. But maybe they will be less likely to doubt their eyes should they, someday, see "lightning" on the Moon or a bright star near the setting Sun!

Appendix A

The Danjon Scale of Lunar Eclipse Brightness

The French astronomer A. Danjon proposed a useful five-point scale for estimating the brightness or darkness of an eclipsed Moon. Estimates made on this scale are very useful for determining the condition of Earth's atmosphere at the time of an eclipse.

The scale is given in "L" values and is as follows:

L = 0	Very dark eclipse. Moon almost invisible, especially during mid eclipse.
L = 1	Dark eclipse with a grey or brownish coloration. Details on the Moon distinguishable only with difficulty.
L = 2	Deep red or rust-colored eclipse. Central shadow is very dark, while outer edge of umbra (deepest cone of shadow) is relatively bright.
L = 3	Brick-red eclipse. Umbral region usually has a bright or yellow rim.
L = 4	Very bright copper-red or orange eclipse. In these eclipses, the umbra has a bluish and very bright rim.

Appendix B

Lunar Eclipses 2011–2050

Date and time	Type of Eclipse (T=total, Pa=partial, Pn=penumbral)
2011 June 15; 20 h 13 min	T
2011 Dec. 10; 14 h 32 min	T
2012 Jun. 4; 11 h 3 min	Pa
2012 Nov. 28; 14 h 33 min	Pn
2013 May 25; 4 h 10 min	Pn
2013 Apr. 25; 20 h 7 min	Pa
2013 Oct. 18; 23 h 50 min	Pn
2014 Apr. 15; 7 h 46 min	T
2014 Oct. 8; 10 h 55 min	T
2015 Apr.4; 12 h	T
2015 Sept. 28; 2 h 47 min	T
2016 Mar. 23; 11 h 47 min	Pn
2016 Aug. 18; 9 h 42 min	Pn
2016 Sept. 16; 18 h 54 min	Pn
2017 Feb. 11; 0 h 4 min	Pn
2017 Aug 7; 18 h 20 min	Pa
2018 Jan. 31; 13 h 30 min	T
2018 Jul. 27; 20 h 22 min	T
2019 Jan. 21; 5 h 12 min	T
2019 Jul. 16; 21 h 31 min	Pa
2020 Jan. 10; 19 h 10 min	Pn
2020 Jun. 5; 19 h 25 min	Pn

Date and time	Type of Eclipse (T = total, Pa = partial, Pn = penumbral)
2020 Jul. 5; 4 h 30 min	Pn
2020 Nov. 30; 9 h 43 min	Pn
2021 May 26; 11 h 19 min	T
2021 Nov. 19; 9 h 3 min	Pa
2022 May 16; 4 h 11 min	T
2022 Nov. 8; 10 h 59 min	T
2023 May 5; 17 h 23 min	Pn
2023 Oct. 28; 20 h 14 min	Pa
2024 Mar. 25; 7 h 13 min	Pn
2024 Sept. 18; 2 h 4 min	Pa
2025 Mar. 14; 6 h 59 min	T
2025 Sept. 7; 18 h 12 min	T
2026 Mar. 3; 11 h 33 min	T
2026 Aug. 28; 4 h 13 min	Pa
2027 Feb. 20; 23 h 13 min	Pn
2027 Jul. 18; 16 h 3 min	Pn
2027 Aug 17; 7 h 14 min	Pn
2028 Jan. 12; 4 h 13 min	Pa
2028 Jul. 6; 18 h 19 min	Pa
2028 Dec. 31; 16 h 52 min	T
2029 Jun. 26; 3 h 22 min	T
2029 Dec. 20; 22 h 42 min	T
2030 Jun. 15; 22 h 33 min	Pa
2030 Dec. 9; 22 h 27 min	Pn
2031 Jun. 5; 11 h 44 min	Pn
2031 May 7; 3 h 51 min	Pn
2031 Oct. 30; 7 h 45 min	Pn
2032 Apr. 25; 15 h 13 min	T
2032 Oct. 18; 19 h 02 min	T
2033 Apr. 14; 19 h 12 min	T
2033 Oct. 8; 10 h 55 min	T
2034 Apr. 3; 19 h 5 min	Pn
2034 Sept. 28; 2 h 46 min	Pa

Date and time	Type of Eclipse (T = total, Pa = partial, Pn = penumbral)
2035 Feb. 22; 9 h 5 min	Pn
2035 Aug. 19; 1 h 11 min	Pa
2036 Feb. 11; 22 h 11 min	T
2036 Aug. 7; 2 h 51 min	T
2037 Jan.31; 14 h	T
2037 Jul. 27; 4 h 8 min	Pa
2038 Jan. 21; 3 h 48 min	Pn
2038 Jun. 17; 2 h 43 min	Pn
2038 Jul. 16; 11 h 34 min	Pn
2038 Dec. 11; 17 h 43 min	Pn
2039 Jun. 6; 18 h 53 min	Pa
2039 Nov. 30; 16 h 55 min	Pa
2040 May 26; 11 h 45 min	T
2040 Nov. 18; 19 h 3 min	T
2041 May 16; 0 h 41 min	Pa
2041 Nov. 8; 4 h 33 min	Pa
2042 Apr. 5; 14 h 28 min	Pn
2042 Sept. 29; 10 h 4 min	Pa
2042 Oct. 28; 19 h 33 min	Pn
2043 Mar. 25; 14 h 30 min	T
2043 Sept. 19; 1 h 50 min	T
2044 Mar. 13; 19 h 37 min	T
2044 Sept. 7; 11 h 19 min	T
2045 Mar. 3; 7 h 42 min	Pn
2045 Aug. 27; 13 h 53 min	Pn
2046 Jan. 22; 13 h 1 min	Pa
2046 Jul. 18; 1 h 4 min	Pa
2047 Jan. 12; 1 h 24 min	T
2047 Jul. 7; 10 h 34 min	T
2048 Jan. 1; 6 h 52 min	T
2048 Jun. 26; 2 h 1min.	Pa
2048 Dec. 20; 6 h 26 min	Pn
2049 May 17; 11 h 25 min	Pn

Date and time	Type of Eclipse (T = total, Pa = partial, Pn = penumbral)
2049 Jun 15; 19 h 12 min	Pn
2049 Nov. 9; 15 h 50 min	Pn
2050 May 6; 22 h 30 min	T
2050 Oct. 30; 3 h 20 min	T

Appendix C

Solar Eclipses 2011–2030

Date and time (day; h:min:s)	Type of Eclipse (T = total, P = partial, A= annular)	Magnitude	Where visible
2011 Jan. 4; 8:51:42	P	0.858	Europe, Africa, c. Asia
2011 Jun. 1; 21:17:18	P	0.601	e. Asia, N. America, Iceland
2011 Jul. 1; 8:39:30	P	0.097	s. Indian Ocean
2011 Nov. 25; 6:39:30	P	0.905	s. Africa, Antarctica, Tasmania, NZ
2012 May 20; 23:53:53	A (5m46s)	0.944	Asia, Pacific N. America
2012 Nov. 13; 22:12:55	T (4m02s)	1.050	Australia, NZ, s. Pacific, S. America
2013 May 10; 0:26:20	A (6m3s)	0.954	Australia, NZ, c. Pacific
2013 Nov. 3; 12:47:36	A/T (1m40s)	1.016	e. Americas, s. Europe, Africa
2014 Apr. 29; 6:4:32	A	0.987	s. India, Australia, Antarctica
2014 Oct. 23; 21:45:39	P	0.811	n. Pacific, N. America
2015 Mar. 20; 9:46:47	T (2m47s)	1.045	Iceland, Europe, n. Africa, n. Asia

Date and time (day; h:min:s)	Type of Eclipse (T = total, P = partial, A= annular)	Magnitude	Where visible
2015 Sept. 13; 6:55:19	P	0.788	s. Africa, s. India, Antarctica
2016 Mar. 9; 1:58:19	T (4m9s)	1.045	e. Asia, Australia, Pacific
2016 Sept. 1; 9:8:2	A (3m6s)	0.974	Africa, Indian Ocean
2017 Feb. 26; 14:54:32	A (44s)	0.992	s. S. America, Atlantic, Africa, Antarctica
2017 Aug. 21; 18:26:40	T (2m40s)	1.031	N. America, n. S. America
2018 Feb. 15; 20:52:33	P	0.599	Antarctica, s. S. America
2018 Jul. 13; 3:2:16	P	0.336	s. Australia
2018 Aug. 11; 9:47:28	P	0.737	n. Europe, ne. Asia
2019 Jan. 6; 1:42:38	P	0.715	ne. Asia, n. Pacific
2019 Jul. 2; 19:24:7	T (4m33s)	1.046	s. Pacific, S. America
2019 Dec. 26; 5:18:53	A (3m39s)	0.970	Asia, Australia
2020 Jun. 21; 6:41:15	A (38s)	0.994	Africa, se. Europe, Asia
2020 Dec. 14; 16:14:39	T (2m10s)	1.025	Pacific, s. S. America, Antarctica
2021 Jun. 10; 10:43:6	A (3m51s)	0.943	n. N. America, Europe, Asia
2021 Dec. 4; 7:34:38	T (1m54s)	1.037	Antarctica, s. Africa, s. Atlantic
2022 Apr. 30; 20:42:36	P	0.640	se. Pacific, s. S. America
2022 Oct. 25; 11:1:19	P	0.862	Europe, ne. Africa, Middle East, w. Africa

Date and time (day; h:min:s)	Type of Eclipse (T = total, P = partial, A= annular)	Magnitude	Where visible
2023 Apr. 20; 4:17:55	A/T (1m16s)	1.013	se. Asia, E. Indies, Australia, Philippines, NZ
2023 Oct. 14; 18:0:40	A (5m17s)	0.952	N., C., & S. America
2024 Apr. 8; 18:18:29	T (4m28s)	1.057	N. & C. America
2024 Oct. 2; 18:46:13	A (7m28s)	0.933	Pacific, s. S. America
2025 Mar. 29; 10:48:36	P	0.938	nw. Africa, Europe, Russia
2025 Sept. 21; 19:43:4	P	0.855	s. Pacific, NZ, Antarctica
2026 Feb. 17; 12:13:5	A (2m20s)	0.963	s. Argentina & Chile, Africa, Antarctica
2026 Aug 12; 17:47:5	T (2m18s)	1.039	n. N. America, w. Africa, Europe
2027 Feb. 6; 16:0:47	A (7m51s)	0.928	S. America, Antarctica, w. & s. Africa
2027 Aug. 2; 10:7:49	T (6m23s)	1.079	Africa, Europe, Middle East
2028 Jan. 26; 15:8:58	A (10m27s)	0.921	e. N. America, C. & S. America, w. Europe, nw. Africa
2028 Jul. 22; 2:56:39	T (5m10s)	1.056	se. Asia, E. Indies, Australia, NZ
2029 Jan. 14; 17:13:47	P	0.871	N. & C. America
2029 Jun. 12; 4:6:13	P	0.458	Arctic, Scandinavia, Alaska, n. Asia, n. Canada
2029 Jul. 11; 15:37:18	P	0.230	s. Chile, s. Argentina
2029 Dec. 5; 15:3:57	P	0.891	s. Argentina, s. Chile, Antarctica
2030 Jun. 1; 6:29:13	A (5m21s)	0.944	Europe, n. Africa, Middle East, Asia, Arctic, Alaska

Date and time (day; h:min:s)	Type of Eclipse (T = total, P = partial, A= annular)	Magnitude	Where visible
2030 Nov. 25; 6:51:37	T (3m4s)	1.047	s. Africa, s. Indian Ocean, E. Indies, Australia, Antarctica

Appendix D

Transits of Mercury 2016–2100

May 9	2016
November 11	2019
November 13	2032
November 7	2039
May 7	2049
November 9	2052
May 10	2062
November 11	2065
November 14	2078
November 7	2085
May 8	2095
November 10	2098

Subject Index

Name Index